THE NEW GROVE®

EARLY KEYBOARD INSTRUMENTS

THE NEW GROVE®

DICTIONARY OF MUSICAL INSTRUMENTS

Editor: Stanley Sadie

The Grove Musical Instruments Series

EARLY KEYBOARD INSTRUMENTS

ORGAN

PIANO

VIOLIN FAMILY

in preparation

BRASS

WOODWIND

THE NEW GROVE®

EARLY KEYBOARD INSTRUMENTS

Edwin M. Ripin Denzil Wraight
G. Grant O'Brien Howard Ferguson
John Caldwell Howard Schott
William Dowd

MACMILLAN

Parts of this material first published in The New Grove® Dictionary of Musical Instruments, edited by Stanley Sadie, 1984

and

The New Grove Dictionary of Music and Musicians®, edited by Stanley Sadie, 1980

First published in UK in paperback with additions 1989 by
PAPERMAC
a division of Macmillan Publishers Limited
London and Basingstoke

First published in UK in hardback with additions 1989 by
MACMILLAN LONDON LIMITED
4 Little Essex Street London WC2R 3LF
and Basingstoke

British Library Cataloguing in Publication Data

Early keyboard instruments. — (The new Grove musical instruments series)
1. Keyboard instruments
I. Ripin, Edwin M. II. The new Grove dictionary of musical instruments
786.2′2 ML549

ISBN 0-333-44449-3 (hardback)
ISBN 0-333-44450-7 (paperback)

First American edition in book form with additions 1989 by
W. W. NORTON & COMPANY

500 Fifth Avenue New York NY 10110

ISBN 0-393-02554-3

ISBN 0-393-30515-5 {PBK.}

Printed in Hong Kong

1 2 3 4 5 6 7 8 9 0

Contents

List of illustrations

Illustration acknowledgments

We are grateful to those listed below for permission to reproduce illustrative material: Bibliothèque Nationale, Paris (figs.2, 39, 50); Oxford University Press (fig.3: from D. H. Boalch, *Makers of the Harpsichord and Clavichord, 1440–1840*, Oxford, 2/1974); Board of Trustees of the Victoria and Albert Museum, London (figs.4, 13, 16, 27, 28, 32, 36, 46, cover); Museo Civico Medievale, Bologna (fig.5); photo G. Grant O'Brien (figs.6, 7, 11); Russell Collection of Harpsichords and Clavichords, University of Edinburgh (figs.8, 17, 25, 34, 44, 45); Staatliches Institut für Musikforschung Preussischer Kulturbesitz, Berlin, (fig.9); photo Thomas A. Brown (fig.10); Trustees of the National Gallery, London (figs.12, 33); Warwickshire Museum Collection, Warwick (fig.15); Gesellschaft der Musikfreunde, Vienna/photo Kunsthistorisches Museum, Vienna (fig.19); Museum für Kunst und Gewerbe, Hamburg (fig.20); Société Pleyel SA, Paris (fig.22); J. C. Neupert, Bamberg and Nuremberg (fig.23); Frank Hubbard Harpsichord Kits Inc., Waltham, Massachusetts (fig.24); Musée Instrumental, Conservatoire Royal de Musique, Brussels (figs.26, 35); Stedelijke Musea, Bruges/photo ACL, Brussels (fig.29); Metropolitan Museum of Art (Gift of B. H. Homan, 1929), New York (fig.30); Germanisches Nationalmuseum, Nuremberg (fig.31); Soprintendenza per i Beni Artistici e Storichi delle Marche, Urbino (fig.40); Worcester Art Museum, Worcester, Massachusetts (fig.41); Musikinstrumenten-Museum, Karl-Marx-Universität, Leipzig (fig.42); Universitätsbibliothek, Marburg (fig.47); Royal College of Music, London (fig.48); The Earl of Wemyss and March/photo Edinburgh University Press (fig.49); Alexandr Buchner, Prague (fig.53); British Library, London (fig.54); Provost and Fellows of King's College, Cambridge (fig.57)

General abbreviations

attrib. — attribution, attributed to

b — born
BWV — Bach-Werke-Verzeichnis [Schmieder, catalogue of J. S. Bach's works]

c — circa [about]
CNRS — Centre Nationale de la Recherche Scientifique

d — died
diss. — dissertation

F — Falck [catalogue numbers in M. Falck, *Wilhelm Friedemann Bach* (Leipzig, 1913, 2/1919)]
facs. — facsimile
fl — floruit [he/she flourished]

Hob. — Hoboken catalogue [Haydn]
hpd — harpsichord
Hz — Hertz [cycles per second]

incl. — includes, including

Jb — Jahrbuch [yearbook]
Jg. — Jahrgang [year of publication/ volume]

kbd — keyboard

n.d. — no date of publication

pf — piano
pll — plates
pubd — published

R — photographic reprint
r — recto
repr. — reprinted
rev. — revision, revised (by/for)

ser. — series
suppl. — supplement

trans. — translation, translated by
transcr. — transcription, transcribed by/for

U. — University

v — verso

WQ — Wotquenne catalogue [C.P.E. Bach]

Bibliographical abbreviations

AcM — *Acta musicologica*
AMf — *Archiv für Musikforschung*
AMP — Antiquitates musicae in Polonia
AMw — *Archiv für Musikwissenschaft*
AnM — *Anuario musical*
AnMc — *Analecta musicologica*
AnnM — *Annales musicologiques*

CaM — Catalogus musicus
CHM — *Collectanea historiae musicae*

DDT — Denkmäler deutscher Tonkunst
DM — Documenta musicologica

FoMRHI Quarterly — *Fellowship of Makers and Restorers of Historical Instruments Quarterly*

GfMKB — *Gesellschaft für Musikforschung Kongressbericht*
Grove 4 — *Grove's Dictionary of Music and Musicians*, 4th edn.
Grove I — *The New Grove Dictionary of Musical Instruments*
Grove 6 — *The New Grove Dictionary of Music and Musicians*
GSJ — *The Galpin Society Journal*

IMSCR — *International Musicological Society Congress Report*

JAMIS — *Journal of the American Musical Instrument Society*
JAMS — *Journal of the American Musicological Society*
JbMP — *Jahrbuch der Musikbibliothek Peters*

KJb — *Kirchenmusikalisches Jahrbuch*

MB — Musica brittanica
MD — *Musica disclipina*
Mf — *Die Musikforschung*
MGG — *Die Musik in Geschichte und Gegenwart*
ML — *Music and Letters*
MMg — *Monatshefte für Musikgeschichte*
MMN — Monumenta musicae neerlandicae
MQ — *The Musical Quarterly*
MR — *The Music Review*
MSD — Musicological Studies and Documents
MT — *The Musical Times*

NBJb	*Neues Beethoven-Jahrbuch*
NOHM	*The New Oxford History of Music*
PMA	*Proceedings of the Musical Association*
PRM	*Polski rocznik muzykologiczny*
PRMA	*Proceedings of the Royal Musical Association*
RaM	*La rassegna musicale*
RBM	*Revue belge de musicologie*
RdM	*Revue de musicologie*
ReM	*La revue musicale*
RMARC	*R[oyal] M[usical] A[ssociation] Research Chronicle*
RMI	*Rivista musicale italiana*
SIMG	*Sammelbände der Internationalen Musik-Gesellschaft*
SMw	*Studien zur Musikwissenschaft*
STMf	*Svensk tidskrift för musikforskning*
TVNM	*Tijdschrift van de Vereniging voor Nederlandse muziekgeschiedenis*
VMw	*Vierteljahrsschrift für Musikwissenschaft*
ZIMG	*Zeitschrift der Internationalen Musik-Gesellschaft*
ZMw	*Zeitschrift für Musikwissenschaft*

Preface

The late Edwin M. Ripin was the original author of most of the articles in *The New Grove Dictionary of Music and Musicians* that have been drawn upon in this book. However, because of his early death (in 1975) and the continuing progress of research, his material was revised by Howard Schott before its first publication and supplemented by material supplied by John Barnes and G. Grant O'Brien. The final section on the harpsichord, since 1800, was written by Dr Schott. In the course of revision for publication in *The New Grove Dictionary of Musical Instruments*, Denzil Wraight provided new sections on the Italian harpsichord in the Renaissance and the 18th century and William Dowd a new section on the French harpsichord in the 18th century (as well as supplying revisions on the French instrument at other periods). For this volume, these authors have supplied further revisions to existing material; further, Mr Wraight has supplied a new text on Italy in the period *c*1590–*c*1700 and G. Grant O'Brien on Flanders in that period (drawing on his *Grove* entry on the Ruckers family).

With the exception of sections (5) and (6), originally written respectively by Howard Schott and Peter Williams, Chapter 4 has been assembled by Denzil Wraight, using as a basis (except for sections (2) and (3), which are his own) material by Edwin M. Ripin. Mr Wraight also supplied Appendices 1 and 2. Chapter 5 is based on Howard Ferguson's entry 'Keyboard Music' written for *The New Grove*, with additional material; Dr Ferguson also supplied material towards Appendix 3, the bulk of which was prepared by John Caldwell, who also supplied the bibliography for this section (based on his entry in *The New Grove*, 'Keyboard Music: bibliography').

S.S.

CHAPTER ONE

The Harpsichord

The harpsichord is a string keyboard instrument, distinguished from the clavichord and the piano by the fact that its strings are plucked rather than struck and characterized by an elongated wing shape like that of a grand piano. As in the grand piano, this shape results from the fact that the strings, growing progressively longer from treble to bass, run directly away from the player, in contrast to the oblique stringing of a spinet and the transverse stringing of a virginal. The earliest known reference to a harpsichord dates from 1397, when a jurist in Padua wrote that a certain Hermann Poll claimed to have invented an instrument called the 'clavicembalum'; and the earliest known representation of a harpsichord is a sculpture in an altarpiece of 1425 from Minden in north-west Germany. Other terms for the instrument include *clavecin* (French); *Cembalo, Kielflügel* (German); *cembalo* and *clavicembalo* (Italian).

The instrument remained in active use up to and throughout the 18th century, not only for the performance of solo keyboard music but also as an essential participant in chamber music, orchestral music and opera; in fact it retained the last of these functions long after most solo keyboard music and chamber music involving a keyboard was being composed with the piano in mind. The harpsichord had almost completely fallen into disuse by about 1810; its modern revival dates from the 1880s. (For a discussion of the repertory see Chapter Five.)

In describing keyboards in this volume the following conventions have been followed: an oblique stroke (e.g. *C/E*) indicates a short octave (i.e. an apparent lowest key *E* is tuned to *C* to extend the compass downwards without increasing the overall dimensions of the instrument); a comma indicates a missing accidental (e.g. *G′,A′* signifies the absence of *G♯′*).

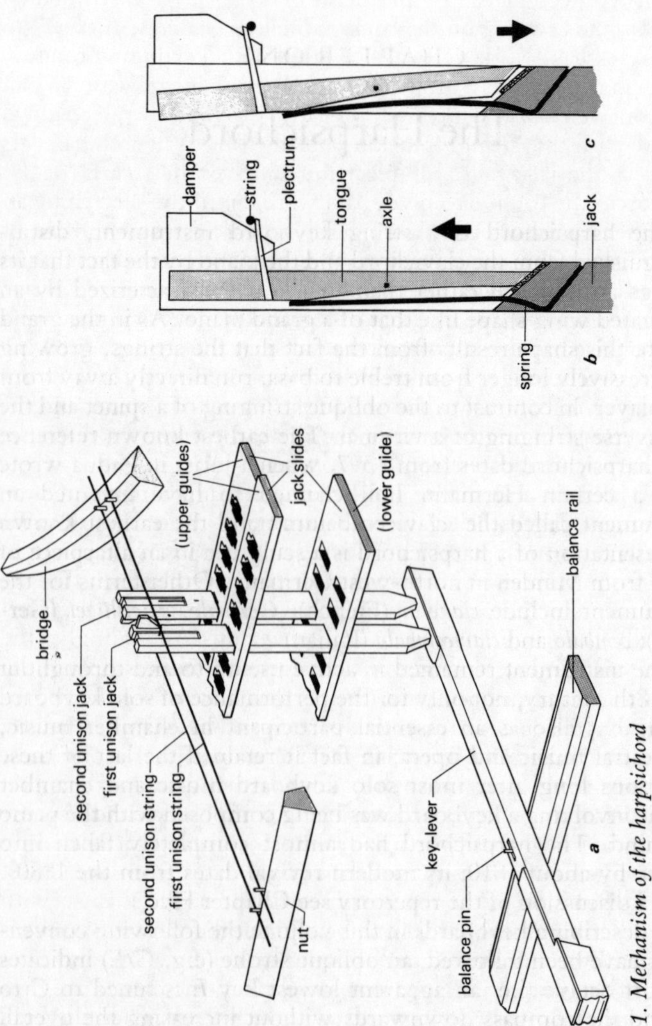

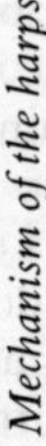
1. Mechanism of the harpsichord

1. STRUCTURE

The heart of the harpsichord's mechanism is the jack, a slender slip of wood (replaced by plastic in many modern instruments) which stands resting on the back of the key (see fig. 1*a*). The top of the jack has a wide vertical slot fitted with a swinging tongue, which in turn carries a plectrum of quill, leather or plastic. When the front of a key is depressed, the jack rises, and the plectrum is forced past the string, plucking it (fig. 1*b*). When the key is released, the jack falls, the plectrum touches the string (fig. 1*c*) and forces the tongue to pivot backwards until the plectrum can pass the string, after which a light spring (formerly made of bristle or thin brass but often now of plastic) returns the tongue forward into its original position. Meanwhile, a piece of soft but stiff cloth held in a slot next to the tongue makes contact with the string, damping its vibrations and silencing it. A padded bar placed overhead – the jackrail – prevents the jack from flying out of the instrument when the key is struck. In many instruments the jackrail alone limits the vertical motion of the jacks and thereby defines the depth of touch.

This elegant and simple mechanism, though capable of producing any degree of legato or detachment of notes with great sensitivity, is incapable of producing any appreciable change in loudness in response to a change in the force with which the key is struck, since, regardless of force, the string is displaced virtually the same amount by the plectrum. Accordingly, the player can produce conspicuous changes in loudness only if the harpsichord is equipped with devices that can change the degree to which the plectrum extends beyond the string (thereby changing the amount the string is displaced when it is plucked), or if each key is provided with additional jacks and strings that the player may engage or disengage at will. The second of these options is much the more important; it is greatly facilitated by the harpsichord's longitudinal stringing, which permits each set of jacks to be placed in a row perpendicular to the strings, with as many rows as desired set one behind another. A set of jacks is engaged by being shifted towards the strings by a lateral movement of the slotted jackslide that supports it; the plectra are thus positioned below the strings and will pluck them when the keys are depressed. A set of jacks is disengaged by shifting the jackslide in the opposite direction so that the plectra pass the strings without plucking them when the keys are depressed.

Although some harpsichords have only a single set of strings and jacks, most have at least two sets with the jacks facing in opposite directions (see fig. 1*a*). (Accordingly, the strings associated with each key are usually widely spaced to permit the jacks to pass between them, and the closely spaced pairs of strings on such a harpsichord are not tuned to the same pitch but, rather, to adjacent notes.) An arrangement of this kind permits two strings associated with a single key to be placed on a single level; but if there are more than two sets of strings, some must pass the jacks at a different level. The use of different levels for different sets of strings is especially associated with the fact that ordinarily no more than two of the sets of strings on a harpsichord are tuned to the same pitch. Rather, an additional set of strings is likely to be tuned an octave above normal pitch; a rare, fourth set is tuned an octave below normal pitch; and a still rarer fifth set of strings is tuned two octaves above normal pitch. (As on organs, normal pitch is termed 8′ pitch, a pitch an octave higher is termed 4′ pitch, a pitch an octave lower 16′ pitch, and a pitch two octaves above 8′ pitch is termed 2′ pitch.)

These higher and lower pitches are best sounded by strings proportionally shorter and longer than those which sound 8′ pitch, and it is most convenient to arrange such strings on their own bridges with the shorter ones at a lower level and the longer ones at a higher level. Thus, on a typical 18th-century harpsichord equipped with two sets of 8′ strings and one set of 4′ strings, the 4′ strings would be at a low level, with the wrest plank (pin block) bridge (nut) placed near the jacks and close to the edge of the wrest plank, and the soundboard bridge placed at an appropriate distance away on the soundboard (see fig. 14, p. 60). The two sets of 8′ strings would both pass over a separate, higher nut placed further from the edge of the wrest plank and a separate, higher bridge placed further back on the soundboard. Similarly, in the unusual harpsichords made in Hamburg by the Hass family, for example, the 16′ strings pass over their own nut and bridge, which are higher than the 8′ nut and bridge and placed outside them, while the short 2′ strings pass over a very low bridge and nut placed between those of the 4′ strings.

Two stringing materials were in wide use throughout the history of the harpsichord: iron and brass. Iron wire was a hard-drawn, pure material (without strengthening carbon, i.e. not 'steel' in the modern sense) and brass was of two types depending on the proportions of zinc and copper: 'yellow brass' (about

25–30% zinc, 75–70% copper) and 'red brass' (about 10–15% zinc, 90–85% copper). Red brass has a lower tensile strength than yellow brass and is used for the lowest few notes in some instruments. More costly materials such as gold and silver are described in some sources, but it is unlikely that these were ever used more than occasionally.

The pitch at which a given string length breaks is practically a constant for each material and substantially independent of diameter, even though it might seem 'intuitively' correct that a thicker string can come to a higher pitch. Since the tensile strength of iron is greater than that of yellow brass, and yellow brass greater than that of red brass, the scales of iron-strung instruments are longer than those of brass-scaled instruments. Thus, two basic groups of scalings arise for all instruments at normal pitch: about 32 to 36 cm at *c''* for iron scales and 25 to 28 cm for brass scales. Both iron and brass scales are found in all instrument-making traditions, although at certain times particular tastes have predominated.

As noted, different levels of strings are required whenever there are more than two sets; accordingly, the relatively rare harpsichords with three sets of 8′ strings must carry them at two different levels where they pass the jacks. This is normally accomplished by using two bridges and two different stringing materials, with two shorter strings of one material (brass) on one bridge, and one longer set of strings (iron) for the other, all tuned to the same pitch. Otherwise either a stepped nut or two separate nuts are used; however, since the separation of levels is required only where the strings pass the jacks, a single bridge without a step may be used on the soundboard. On instruments with a single set of 8′ and a single set of 4′ strings, each set passes over its own bridge and nut, with the 4′ strings on a lower level.

The position of the tuning pins and the hitch-pins for the 4′ strings raises difficulties, since if they were placed with those of the 8′ strings (in the front part of the wrest plank and in the case lining respectively) the 4′ strings would have to pass through the 8′ nut and bridge and there would, in addition, be an inordinate length of unused 4′ string beyond the 4′ bridge, which would tend to make the 4′ strings go out of tune easily. Accordingly, the tuning pins for the 4′ strings are usually placed between the 8′ and 4′ nuts. Italian 16th-century harpsichords, however, always had the 8′ and 4′ tuning pins together. 4′ hitch-pins are driven into the soundboard between the 4′ and the 8′ bridges. A strengthening bar or 4′ hitch-pin rail is usually glued to the underside of the soundboard to withstand the tension exerted on the 4′ hitch-pins by the strings. This bar also in effect divides the soundboard into two distinct areas, one of which, lying between the 4′ hitch-pin rail and the curved side of the case, serves the 8′ strings while the other, between the 4′ hitch-pin rail and (usually) an oblique cut-off bar, serves the 4′ strings. The triangular area of the soundboard to the left of the cut-off bar is generally stiffened by the number of transverse ribs. (If 2′ strings are present, they are hitched to pins driven through the soundboard into a second hitch-pin rail, placed between the 2′ and 4′ bridges. If 16′ strings are present, the 8′ strings may pass through holes in the 16′ nut and bridge, or the 16′ strings, instead of having a bridge of their own, may be carried by the 8′ bridge; if the 16′ strings have their own bridge, it may rest on a separate soundboard, and in this event a curving rail separating the two soundboards acts as a hitch-pin rail for the 8′ strings.)

A harpsichord case consists of five basic parts. Clockwise from the left, these are: the spine, the long straight side of the case at the player's left; the tail, a short straight piece set at an acute angle to the spine; the bentside, a curving section that runs more or less parallel to the bridge (occasionally the bentside and tail are combined in a single S-shaped piece, yielding a case with a curved tail rather like that of a modern grand piano); the cheekpiece, a short straight piece at the player's right; and the bottom, which on all harpsichords from the 16th century to the 18th is a piece of wood that closes the instrument and thereby performs both a structural and an acoustical function. The wrest plank is set between the cheekpiece and the spine, with space below it for the keyboard. A space is created for the jackslides between the wrest plank and the belly rail (or header), a transverse member which is sometimes divided into separate upper and lower parts. In this case the lower part is set somewhat behind the upper one in order to leave room for the keys to extend beyond the jacks and reach the slotted rack by which they are usually guided at the back. The upper surface of the belly rail supports the front edge of the soundboard, the other edges of which rest on liners glued to the inside of the spine, tail, bentside and cheekpiece; the 8′ hitch-pins are driven into the liners along the tail and bentside.

It is the case of the instrument which must be capable of resisting the tension of the strings. Although the total string tension in a harpsichord is substantially less than that in a piano, where two or three strings are provided for each note, it is nevertheless a considerable load for a wooden structure. Sometimes the design for the case has a weakness, but in many instances instruments were restrung in the 19th century more heavily than was desirable. The results – warped cases, wrenched-out wrest planks and collapsed soundboards – can be seen in many instrument collections.

The methods by which the cases of historical instruments are braced to withstand the tension of the strings are numerous; apart from those remarkable instruments that have survived with no framing at all, they may be divided into three groups. In the simplest system, a series of vertical braces running nearly the entire depth of the case somewhat beneath the soundboard are placed transversely between the bentside and the spine, and/or obliquely between the bentside and the spine and the bentside and the belly rail; these braces may be supplemented by slanting bars (struts) running downwards from the bentside liner to the bottom. In the second system triangular knees (blocks) are set between the sides and the bottom of the case and between the belly rail and the bottom; again, these knees may be supplemented by slanting bars. In the third and most complicated system, there is more than one set of braces: at the bottom of the case running transversely from the bentside to the spine, at a higher level running obliquely from the bentside liner to the spine liner and the belly rail, and even, in 18th-century English instruments, from the upper belly rail to the baseboard (running parallel to the spine).

Nevertheless, in any system the function of the bracing is the same, namely, to prevent the bentside and tail from collapsing inwards under the pull of the strings. A related problem is that the wrest plank is pulled towards the soundboard. It may bow in the middle, but it is more likely that the treble end will be twisted: the edge nearest the keyboard comes up and the edge nearest the jacks is turned down, an effect particularly noticeable in Kirckman harpsichords. The problem arises because the bentside is a relatively weak member: the straighter the bentside the stronger it is, but in a large, five-octave instrument a considerable curve in the treble is inevitable. Under the tension of the strings this curve is pulled towards the soundboard. Since the wrest plank has relatively little torsional stiffness it cannot resist the twisting of the cheekpiece, to which it is attached. This

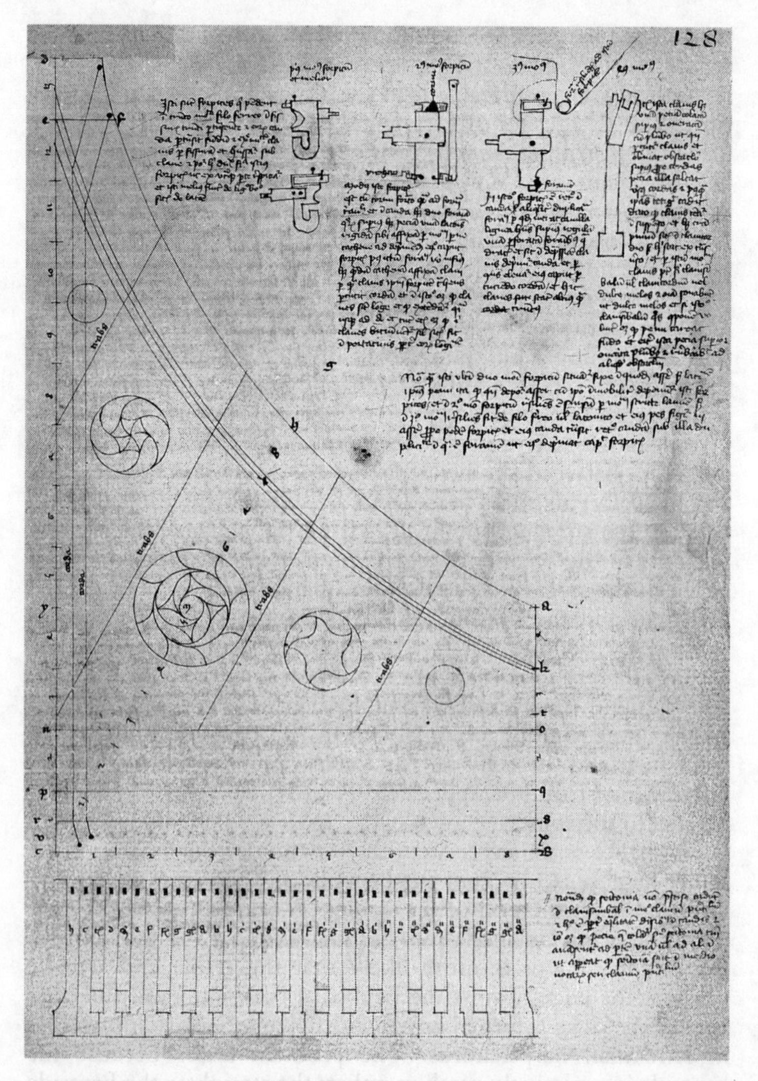

2. Henri Arnaut de Zwolle's plan drawing of a harpsichord (c1440), with descriptions of three types of mechanism, plus one (extreme right) for the dulce melos (see Chapter Four, §8)

twist is exacerbated if the baseboards (bottom boards) are made in two parts (as in many harpsichords) with planks running across the instrument up to the belly rail, and thereafter running the length of the instrument. Since the baseboard contributes to the stiffness of the case, a joint in the baseboard under the belly rail creates a weakness at a critical point. It is obvious from this analysis that the bracing of the bentside, particularly in the treble curve, is extremely important. A further consequence of case weakness is that the soundboard is subjected to considerable shear stresses in the treble area, with the result that a crack often develops close to the cheekpiece.

2. THE RENAISSANCE

15th-century representations of harpsichords from various parts of western Europe generally show short instruments with thick cases. With one exception (which may well have suffered from over-painting) none shows a jackrail, indicating that these instruments did not work by means of jacks as described above. Instead they appear to have had a variety of far more complex actions, such as those cryptically described and illustrated in the manuscript treatise (*c*1440; Bibliothèque Nationale, Paris, MS lat.7295, f.128*r*) of Henri Arnaut de Zwolle (see fig.2).

(i) Italy

Although there are no surviving 15th-century Italian harpsichords, it is known that harpsichords were built in Italy as well as elsewhere in Europe at this time. Thus, any examination of harpsichord making in Italy must rely on illustrations and documents, the evidence of music written for keyboard instruments, and possibly comparisons with other types of instrument. The fruits of such a study are necessarily rather speculative. Manuscripts of Italian keyboard music (including organ music) show that a compass of *F,G,A* to *g''*,*a''* (or *c'''*, or *f'''*) was in use in the second half of the 15th century. An intarsia representation of a harpsichord in the choir stalls of S Lorenzo, Genoa, shows a single-register instrument with a compass of *F,G,A* to *g''*. In view of the early date of the intarsia (*c*1520) and the range of the compass, it seems plausible to assume that it represents the type of harpsichord made in the late 15th century.

The earliest surviving harpsichords and the virginals show signs of having benefited from a long period of development. A broad similarity in details of casework, keyboards and mechanism lends conviction to the idea that instruments of the late 15th century were probably not vastly different from those of the beginning of the 16th century. Unfortunately, the somewhat crudely-made intarsia does not yield much information about constructional details: the case is thick, without mouldings, and the key covers are of boxwood, but most striking is the curious bentside made with two curves, an apparently unique example among harpsichords (illustrated in Winternitz, 1967).

Since about 40 dated harpsichords have survived from before 1600, instrument construction in the 16th century can be described in detail. Cases were usually thin (3–6 mm) and made of cypress, although maple was occasionally used. The slender shape of the case, which is characteristically Italian, results from the practice of doubling the string length at each octave down to the tenor register; the case sides were usually not so deep as in instruments from other countries. Mouldings at the top and bottom of the case contribute to an appearance of architectural elegance. Such thin-cased instruments were rarely painted, but were provided with a separate, decorated case. (This arrangement of an instrument with an outer case is often referred to as an 'inner-outer' instrument.) Stands upon which the outer case was supported survive in a variety of forms, from simple, turned baluster legs to carved stands that were painted and gilded. An exceptional specimen (in the Metropolitan Museum of Art, New York) has gilded, fish-tailed tritons, sea-nymphs and dolphins, all of which combine to support the instrument (see Winternitz, 1961).

Some harpsichords with a single 8′ register have survived without internal bracing – the limited amount of tension of one register made this possible. Internal bracing usually comprised two to three stiffening rails glued to the bottom boards; triangular blocks (knees) then maintained the sides perpendicular to the bottom board. Two or three knees on the spine side and five to seven on the bentside was a common arrangement. Diagonal struts from the bentside liner to the bottom were also used, either in cooperation with knees or as the only method of supporting the sides, although the use of diagonal struts alone is known to occur only in Venetian harpsichords.

The cheeks at either side of the keyboard were often cut to a combination of scroll, half circle and other shapes. Decorative inlaid stripes of contrasting colours in geometrical designs, painted arabesques and patterns, and small ivory knobs were used to embellish some of the finer instruments. The natural keys were usually covered with boxwood, sometimes with ivory; only rarely were dark woods used. Sharp keys were commonly made of black-stained wood topped with ebony veneer or sometimes with stripes of different types of wood. The nameboard was sometimes panelled with mouldings, and, if signed, often carried the maker's name in small capital letters. Many instruments are without date or signature. The soundboards often had a rose consisting of several layers of parchment and/or thin wood veneer of gothic or interweaving geometric design. Some instruments had three or four roses, a feature which echoes the illustrations of 15th-century European harpsichords and Arnaut de Zwolle's manuscript (see fig.2).

Keyboards were most frequently made of quartered beech, but in a few instruments lime, walnut, chestnut or pine was used. The jack end of the key-lever was guided by a wooden tongue in a vertical slot on the rack. The keys were arrested by cloth padding on the front key-frame rail, or by the jacks reaching the padded underside of the jackrail, or perhaps a combination of the two. Since no unaltered instruments survive, it is impossible to know exactly how the touch was regulated. The amount of sharp projecting above the natural-key covers indicates a fairly shallow depth of touch (about 6 mm) in many instruments.

Italian jacks were normally thicker than those used in other countries, in order to compensate for their short length and thereby ensure a reliable return. They were most often made of pear (or perhaps service wood; in such cases only microscopic analysis can distinguish between closely similar woods). Small springs of flat brass strip were used; the tongue was often centrally placed to enable a damper-slot to be cut on either side. Quill was the usual material employed for the plectra; it cannot be substantiated that leather plectra, which are now found in some jacks, were used originally since many such jacks are either not original or have been altered. A one-piece boxslide about 2.5 to 5 cm deep was used by most Italian makers in preference to the separate upper and lower guides found in other traditions. The jackslides were made either by sawing out the slots and then closing the open side with a thin strip of wood, or by gluing small blocks of wood to a thin strip, with the correct spacing for the thickness of the jacks, and then gluing another strip on the open side. Most 1 × 8′, 1 × 4′ harpsichords were provided with small knobs fixed in the ends of the jackslides and projecting through the cheek, to enable the registers to be engaged and disengaged (see also §3(i) below). The jackrail was often decorated with mouldings, either cut into the wood or fixed on to the rail. To hold it in place, slotted blocks were glued to the inside of the case; the rail is slid towards the tail for removal.

A large number of 16th-century harpsichords have cypress (*Cupressus sempervirens*) soundboards, but spruce (of the genus *Picea*) and what in many instances may be fir (of the genus *Abies*) were also used. Cypress, walnut and beech were normally used for bridges, which typically had a smaller cross-section than those found in other countries. They were usually made from parallel-sided stock (in contrast to the often basically triangular shape of bridges of other traditions), with a moulding cut on the top edge. A reduction in the height and width of the bridge towards the treble was common; sometimes the ends were finished with a decorative scroll. In other countries the bridge often had a tight curve in the bass, but in Italian harpsichords a small piece was usually mitred at an angle to the main bridge for the last few bass strings. Nuts were often of the same material and finished to the same dimensions and with the same decoration as the bridge, a feature of construction which has enabled many later alterations to be detected. They were either fixed on a straight line on the wrest plank or curved, with the inside of the curve facing the jacks. (Nuts with the curve in the opposite direction can usually be shown to be later alterations, which shortened the scaling.)

Ribbing systems have been found with three or four bars ('cross bars') running at an angle from the spine (towards the front of the instrument) and crossing under the bridge (where they are sometimes undercut to leave the soundboard free). Others have a cut-off bar, with or without additional cross bars. Some harpsichords seem to have been made without any bars at all. The impossibility of gaining access to the inside of many instruments makes it difficult to establish how rigidly makers followed these systems, and, as with instruments of other countries, exceptions to these types of barring can be found. These systems of barring are found in Italian harpsichords from the 16th to the 18th centuries, so that no feature can be categorically assigned to one period of instrument making. Furthermore, the acoustical function of barring on harpsichords is not well enough understood for specific conclusions to be drawn about the sound of a harpsichord simply from the type of barring used.

Where a 4′ stop was part of the specification, the 4′ hitch-pins were often simply driven into the soundboard and secured on the underside with a drop of glue. This practice is only possible with a relatively hard wood such as cypress (and not with spruce or fir), but thin 4′ hitch-pin rails glued to the soundboard were also used. The 8′ strings were fixed to the soundboard liner in the conventional way, and some later harpsichords have an extra hitch-pin rail on the soundboard to enable double-pinning to be used in the bass.

The point at which the string is plucked is important in determining the character of the sound of an instrument. When the plucking point is near the nut ('close plucking'), the sound is nasal; nearer the middle of the string ('centre plucking'), it is rounder. It can be seen in Italian harpsichords that the plucking point (in the back register, furthest from the player) for an 8′ lay close to a third of the string length in most instances at *c′′*. At the extreme treble the plucking point was nearer the middle of the string. In the bass the plucking point was, for obvious practical reasons of key length, relatively close to the nut.

A common arrangement of the plucking directions for a 2 × 8′ disposition would have one 8′ plucked to the left (as seen by the player) and the other, with the jacks nearer the keys, to the right. Some dispositions would appear to have been originally in the reverse order, but repairs and alterations often make it impossible to establish such matters beyond doubt. The difference in sound of the two configurations when both 8′ strings are played is small.

In the case of the 1 × 8′, 1 × 4′ disposition it can be clearly shown from several examples that the registers were arranged with the 8′ plucked to the left and the 4′ (with the jacks nearer the player) to the right. A comparison between this disposition and that of Ruckers harpsichords is interesting as it reveals a basic difference in the design, and hence the sound: Ruckers instruments have the 4′ plucked to the left and the 8′ (jacks nearer the player) to the right. This gives a more nasal character to the 8′ than in the Italian harpsichord.

Given the scales and plucking points that the Italian makers chose, the two nuts often lie quite close to each other, making it impracticable to locate the 4′ tuning pins between the 4′ and 8′ nuts (which is the commonest practice in harpsichord making). Instead, holes were drilled through the 8′ nut so that the 4′ strings could reach their tuning pins at the edge of the wrest plank. One harpsichord which has retained its original 4′ in use and still has this layout of the tuning pins is the 1561 Franciscus Patavinus in the Deutsches Museum, Munich.

Many harpsichords had the line of jacks running not at 90° to the long side, but at a slight angle to it so that the jackslides are nearer the front of the instrument at the treble end. This reduces the amount of curve in the bentside if other factors of scaling and plucking points are unchanged, which may have been the makers' reason for this arrangement.

Since many 16th-century harpsichords have had their dispositions changed and their soundboards repaired, reinforced, or differently barred, it is hard to tell from such surviving examples how they might originally have sounded. Furthermore, it is not known precisely how much allowance should be made for the ageing of wood or for cracks in the soundboard, so that it is difficult to estimate even roughly how much the sound character of an instrument may have changed. Many 16th-century harpsichords had their 1 × 8′, 1 × 4′ dispositions changed to 2 × 8′ registers (see below) and were re-strung with brass wire instead of iron wire. If the soundboard is left unchanged and an instrument is re-strung with brass wire (with a corresponding allowance for the lower pitch that brass will come to) the sound is noticeably less brilliant, because it dies away more quickly. Thus, many 16th-century harpsichords cannot be heard in their original form, but in the way that they were rebuilt in order to conform to 17th- or 18th-century requirements. Without doubt, this has helped to bring about the view that Italian harpsichords were built in an almost identical fashion for nearly 150 years. It has therefore to be realized that many confident-sounding judgments as to how these harpsichords originally sounded and comparisons of stylistic differences in harpsichord making in various parts of the Italian peninsular are based on the most slender and unreliable evidence.

It is commonplace to attribute to the old makers certain skills in manipulating the dimensions and materials in order to achieve a desired type of sound. Violin making has, in this respect, collected a large body of myths about timber and varnishes.

Whether the harpsichord makers were actually in possession of the skills sometimes imputed to them has never been established. Organological and acoustical research must determine exactly what their capabilities were. Armed with such an understanding it might be possible to perceive the extent to which it is sensible to speak of a 'typical' Italian harpsichord sound, and the extent to which the instrument maker could tailor the sound to meet the demands of an aesthetic ideal.

One example of an atypical instrument is the Dominicus Pisaurensis (Domenico da Pesaro) harpsichord of 1533 (in the Musikinstrumenten-Museum, Karl-Marx-Universität, Leipzig), which has a soundboard extending to the wrest plank, so that both the nut and the bridge are on it. The similarity of this arrangement to the Hans Müller harpsichord of 1537, made in Leipzig, and the Lodewijk Theewes instrument of 1579, made in London, both of which also have a soundboard under the nut, may indicate that this type of harpsichord was more widely known in the 16th century than the small number of such surviving instruments suggests.

A discussion of compasses cannot begin without an appreciation of the extent to which alterations obscure the original condition of many instruments. Barnes (in Ripin, 1971) has drawn attention to some such alterations. Indeed, so often were instruments altered (sometimes so skilfully as to leave little trace of the work), that it may be true to say there survives no dated Italian harpsichord of the 16th century that has not had its compass, disposition or scale altered. This point can hardly be overstressed since most of the information published in easily accessible works (e.g. Hubbard, 1965; Russell, 1959) refers to the altered state of instruments.

It is often stated that Italian harpsichords usually had two 8′ registers, but in many 16th-century instruments a second register has been added to what was originally a 1 × 8′ disposition. A harpsichord which Hubbard (1965, p.5) introduced as a typical Italian instrument, and which is usually cited as being the oldest harpsichord in existence, is that of 1521 by Hieronymus Bononiensis (Victoria and Albert Museum, London). It now has a 2 × 8′ disposition and compass *C/E–d‴*. According to Debenham (1978; see also Schott, *Victoria and Albert Museum: Catalogue*, 1985, p.15), it was originally a 1 × 8′ instrument with 50 strings. Its original compass was probably *C/E* to *f‴*. There is in fact a slightly older harpsichord in

3. Harpsichord by Vincentius, 1515–16

existence which was made by Vincentius (see Wraight, 'Vicentius and the Earliest Harpsichords', 1986; and fig.3). On the underside of the soundboard is a short inscription indicating that the harpsichord was started on the 18th of September 1515. The nameboard (now lost) bore the date 10 February 1516, which was presumably the date of completion (or perhaps delivery). Unfortunately, many clues as to the original compass or scale are missing since so much of the instrument has been altered, but the available evidence suggests that it originally had a 1 × 8′ disposition, and a compass of *C/E* to *f‴*.

Some 16th-century harpsichords with a single 8′ register were recorded earlier and others have since come to light. It has also become clearer that the 2 × 8′ specification was not common in the 16th century. The earliest dated 2 × 8′ harpsichord was built by Dominicus Pisaurensis in 1570 (see Wraight, 'Nouvelles études', 1985) and it seems likely that the four 'gravicembali doppi' used at the *intermedi* performed in Florence in 1565 for the marriage of Francesco I de' Medici were 2 × 8′ instruments, 'doppi' distinguishing them from 1 × 8′ instruments. Since there are no clear criteria for distinguishing 16th-century harpsichords from those of the early 17th century, there is a difficulty in sifting some of the evidence of undated instruments made around the turn of the century. Although the distinction by centuries is quite arbitrary, from about 1600 onwards more 2 × 8′ harpsichords are found, which clearly suggests that a change in their use occurred then.

Several instruments which now have 2 × 8′ disposition were originally 1 × 8′, 1 × 4′. Doubt has been expressed as to whether any 16th-century Italian instruments originally had a 4′ stop (see Thomas, 1980). However, it is possible to be clear on this point, as some makers marked an outline plan on the baseboards when they started to build the harpsichord, indicating the position of the 8′ and 4′ nuts. The 1574 Baffo (fig.4, p.18) is one such harpsichord with a 4′ which underwent such modifications: the disposition was changed to 2 × 8′ and the compass from *C/E* to *f‴* to *G′/B′* to *c‴*. It also had about 20 cm removed from the bass end of the case. The surviving 16th-century 1 × 8′, 1 × 4′ harpsichords are about twice as numerous as the 1 × 8′ instrument, and there are only a few which originally had a 2 × 8′ disposition.

Some small harpsichords survive with two unison registers; these were 4′ instruments, analogous to octave virginals.

4. Harpsichord by Giovanni Baffo, Venice, 1574

Important examples were made by Dominicus Pisaurensis who worked in Venice: they are dated 1543 (Musée Instrumental, Paris Conservatoire) and 1546 (Gesellschaft der Musikfreunde, Vienna).

The earliest 16th-century harpsichords probably had a compass of *F,G,A* to *a''* (sometimes without *g♯''*). Later modifications have obscured the fact that some early keyboards only reached *a''*, and such keyboards do not appear to have been made in Italy after the 1580s. *C/E* to *f'''* was a more common compass than *C/E* to *c'''* in the 16th century, though it may seem surprising that the larger compass preceded the smaller one. The 'arpitarrone' described by Banchieri (2/1611) had a compass of *C,D* to *e''*. A few surviving instruments confirm the use of this type of bass octave. It seems unlikely that instruments were ever made with a compass of *G'/B'* to *c'''* in the 16th century, though several harpsichords were altered to have this range. A compass of *G'/A'* to *c'''* was common towards the end of the 17th century, and the earliest dated example is known from the 1630s (the last digit of the date is illegible). The so-called '1605 Celestini' harpsichord (Gemeentemuseum, The Hague) has a range of *G',A'* to *f'''*, but the inscription is a fake; the instrument was not made by Celestini and must be dated later (see Wraight, 1987). However, the inventory of Medici instruments (Gai, 1969) lists a harpsichord of 1538 made by Dominicus Pisaurensis with a compass of *G',A'* to *a''*; there seems a good chance that at the date of the inventory the harpsichord had not been rebuilt from some other compass. Another instrument with an exceptionally wide range is the 1579 Baffo harpsichord (in the Paris Conservatoire), which originally had a five-octave (57-note) range of *C/E* to *c''''*. This instrument demonstrates some of the difficulties the organologist has to deal with: the instrument was modified twice and the compass is now *A'* to *f'''*, but all of the original keys have survived.

About half of the harpsichords attributable to the 16th century were made in Venice. A clear enough pattern emerges of harpsichords with a 1 × 8', 1 × 4' disposition to justify calling this a Venetian tradition, but this is deceptive, since such instruments were also made elsewhere. It should be noted that information can be gained only from surviving instruments; it is to be hoped that these are a representative sample of what was made. To a small extent inventories provide information, but even here there are problems which can be solved only by first-hand knowledge of the instruments since descriptions often require interpretation.

It would seem that Venice was the most important harpsichord-making centre in 16th-century Italy; that the reputation of Venetian makers was considerable can be seen from the number of instruments that found their way to other parts of Italy and to other countries. Alfonso II of Ferrara had at least six Venetian harpsichords, and Raimund Fugger in Augsburg had five.

The question of the pitch of Italian harpsichords is complicated because no agreement has been reached on some basic matters of stringing. The pitch problem is important not only for the musicological issue of assigning the harpsichord its place in musical life. As already stated (see §1 above), the type of stringing material used can substantially change the musical character of a harpsichord. Italian string lengths covered a wide range from about 15 cm to 42 cm measured at *c''*. It is this wide range of scales (in comparison with those of other instrument-making traditions) which has suggested that different pitches were in use. At the shorter end of the range were octave virginals and harpsichords (virginals have usually been included in these analyses since they belong to the same family of instruments); at the longer end were harpsichords of unusually low pitch. Scales in the middle and lower end of the range have engaged most attention: these varied from about 25 cm to about 36 cm at *c''*.

Some earlier writers believed that a string length could not be identified with a particular pitch (see Russell, 1959, p.32; Hubbard, 1965, p.9). However, more recent evidence of the scalings used for different types of string material (i.e. iron, yellow brass and red brass wire) indicates that there was a wide measure of agreement among harpsichord makers on certain basic principles. That is, that the strength differences between, for example, iron and brass wire would be reflected in a difference between string lengths (see O'Brien, 1981). Other research revealed the accuracy with which Italian harpsichord makers worked when laying out the positions of the bridges (see Wraight, in *Girolamo Frescobaldi*, 1986). Thus it is clear that the pitch relationships between various instruments were reflected in different string lengths. In principle, an examination of the original string lengths, with due consideration of which scalings were intended for iron wire and which for brass, would yield the scheme of pitches in use in the 16th century.

Such attempts have already been made. After examining 33

Italian harpsichords and virginals, Shortridge (1960) attempted to show that certain scales occurred with such regularity that two groups could be discerned. He gave the averages of these two groups as 26.6 and 32.7 cm at *c''*. It was suggested that the correlation of the shorter scales with keyboards of *C/E* to *c'''* range and longer scales with keyboards of *C/E* to *f'''* range indicated that the *C/E* to *f'''* instruments were at a pitch a 4th lower. The notes *f'''* and *c'''* on this interpretation actually sounded at the same frequency. What made this idea seem plausible was the analogy with the Ruckers 'transposing' harpsichords (see §3(i) below) which had keyboards at pitches a 4th apart. Similar arguments were advanced by Barnes (1965) to support the idea that Italian instruments were at two different pitches in order to facilitate transposition.

An essential weakness of these arguments is that there is no exact correlation between the range of the keyboard and whether the scale is long or short, as examination of other Italian instruments has revealed. Furthermore, the difference between the scalings reported is somewhat less than a 4th and therefore cannot confirm the hypothesis. Although Shortridge examined 33 instruments, this is only about a fifth of the total number that have survived from before 1650. Shortridge's list also gives the scales of some altered instruments which further distort the scheme of original pitches. Thus, the scale averages reported by Shortridge are unreliable as a basis for further study.

Nevertheless, this 'transposition theory' still has a certain immediate appeal and has since been repeated (e.g. Haase and Krickeberg, 1981). This is due partly to the transpositions of a 4th, which are known to have taken place, and also to the fact that notes as high as *f'''* were practically uncalled for in printed 16th-century keyboard music. This could suggest that transposition might be a part of the explanation for the two different keyboards. With the idea of transposition in mind, the difference in pitch of a 4th between the *c'''* and *f'''* notes could then appear to be strong circumstantial evidence for a pitch difference of a 4th between the *C/E* to *c'''* and *C/E* to *f'''* keyboards.

However, the more the organological evidence is examined, the less plausible it is to assume that there were two basic pitches for instruments, a 4th apart. Indeed, within the range of the examples of the longer scales it seems possible to distinguish several specific pitches, so that a simple statistical average of several string lengths obscures the problem.

Although the first writers on the problems of scalings assumed that all the instruments were strung with the same type of material (see Shortridge, 1960; Barnes, 1965; Hubbard, 1965), Thomas and Rhodes (1967) argued that the difference between groups of string lengths was simply due to the use of iron and brass scales. While this contribution was important for drawing attention to the use of iron scales in Italian instruments, several difficulties remain in showing which instruments were designed to use iron wire. A contrary view (that all Italian instruments were strung with brass wire, with the pitches proportional to string lengths) was advanced by Barnes (1968; in Ripin, 1971; 1973) and defended against the views of Thomas and Rhodes.

The documentary evidence indicates that some harpsichords and virginals were designed to be strung with iron wire, while others were strung with brass. It is not possible simply to identify a type of stringing material with one type of instrument, even though some evidence has been interpreted as showing this (see Hellwig, 1976).

Since the original makers' principles of stringing have not been completely researched, this outline of the original pitch schemes cannot be definitive. It does seem however that there were several different pitches in use. Some were high: octave and the occasional use of quint pitch; a few were demonstrably low, and would have been a 4th or 5th lower than some other pitches. Whether these relationships are significant for understanding transpositional practices is not yet clear. Others in the middle range (about 28 cm at *c''* and strung with brass wire) would have been close to *a'* = 415 Hz. Italian virginals were usually strung with iron wire and their range of pitches was substantially the same as those of the short-scaled harpsichords strung with brass, although a few virginals were designed for brass scaling. The arguments concerning the long-scaled harpsichords (about 36 cm at *c''*) are complicated, but it seems that these were strung with iron wire and therefore came to a pitch similar to those of the virginals and short-scaled harpsichords.

According to Wraight's interpretation, there was no change of pitch of about a 4th from the 16th century to the 17th, as had been suggested by Barnes. Barnes's argument (apparently also endorsed by Mendel, 1978) that a wide variety of pitches was used, amounting in effect to chaos, can be seen with hindsight to have considered too few of the original scalings to gain a good overall picture of the situation. Pitch levels stayed fairly constant in the 16th century, but rose by about a semitone or tone during the later 17th century and the 18th.

In retrospect, it can be seen that the initial attempts at making sense of the complexities of Italian string lengths grappled with two problems: the organological evidence of pitches, and the musicological question of 'transposition' practices. Since these first studies were undertaken it has become clearer that there were not simply two groups of pitches a 4th apart. It is clear though that transpositions of a 4th or 5th were routine keyboard practices in the 16th century (see Tagliavini, 1974; Meeùs, 1977). As some harpsichords were at low pitches, questions about transposition similar to those posed by Shortridge (1960) and Barnes (1965) must again be asked: whether these harpsichords were 'transposing' instruments depends not simply on a technical explanation of the pitch but on an understanding of their original musical use. It is no less important to understand the use of the various instruments at other pitches (e.g. octave and quint higher). Such considerations are not isolated from organology, since, as has been seen, they are essential concepts used in organizing evidence into comprehensible schemes.

Theorists and inventors were also applying their ideas to produce unusual instruments. In 1548 Dominicus Pisaurensis built a harpsichord for Zarlino with 19 notes to the octave (see Zarlino, 1558), Lindley suggested (*Grove 6*, 'Temperaments', p.666) that is it not certain whether this was intended to be tuned in equal microtones, or to provide sharps not normally available in mean-tone temperaments. Although it is known that some organs of the 15th century had split sharps, it is not clear how often 16th-century harpsichords were provided with them; most of the surviving instruments with split sharps are of 17th-century origin. One harpsichord made for the court of Alfonso II of Ferrara was altered several times and the present keyboard is not original, but the original compass has been determined from the markings on the baseboard: *C,D,E* to *c'''* with five split sharps for *d♯* and *a♭* (see Wraight, 1983). This instrument could not have been built later than 1597, the year of Alfonso's death.

Nicola Vicentino, in defence of his views of musical theory, published *L'antica musica . . . con l'inventione di uno nuovo stromento* (1555), the invention of the 'new instrument' being a harpsichord with 36 keys to the octave so that pure intervals were available to the player instead of the impure intervals

5. Keyboard of the enharmonic harpsichord by Vito Trasuntino, Venice, 1606

necessary in any temperament. Vito Trasuntino, who was apparently one of the most revered 16th-century instrument makers (he built organs as well), left an example of a similar, fully enharmonic instrument in the 'Clavemusicum omnitonum' (Museo Civico Medievale, Bologna; see fig.5).

(ii) Northern Europe

The overwhelming majority of the surviving 16th-century instruments are Italian, and knowledge of harpsichord making elsewhere in Europe before about 1590 is derived largely from scanty written sources and the few surviving instruments made during this period. Of the small number of known stringed keyboard instruments made north of the Alps before 1590, one is a clavicytherium and only three are harpsichords, the others being virginals or clavichords. Similarly, no 16th-century representation of a north European harpsichord is known at present. The oldest known surviving plucked keyboard instrument is the small clavicytherium (*c*1480) in the museum of the Royal College of Music, London. It seems likely that it was made in Ulm. It has a number of features normally associated with Italian instruments, including very thin case sides, but it is clearly not Italian in its origins. (For a discussion of this instrument see Chapter Four, §4; see also fig.48, p.179.)

From the 1566 Fugger inventory (see Smith, 1980) information can be gleaned about some of the harpsichords that were made but have not survived: a leather-covered harpsichord made in England with several registers; a German harpsichord made in Cologne with two keyboards so that two people could play together; and a Flemish harpsichord that took the latter combination a step further, since it was made with four keyboards to enable four people to play four different parts at once.

It is uncertain whether the style of instrument making in Italy around 1500 influenced that of other countries, or whether a similar style simply arose in several different countries at once. The paucity of information makes it impossible to establish the facts one way or another and difficult even to suggest which explanation is the most likely. A harpsichord made by Hans Müller in Leipzig in 1537 (in the Museo degli Strumenti Musicali, Rome; see fig.6, p.26) is similar in its basic case construction to the thin-cased Italian harpsichord. However, with its massive mouldings and its relatively short length, it would not be easily mistaken for an Italian instrument. Some details of construction

6. Plan and side views of harpsichord by Hans Müller, Leipzig, 1537

further distinguish it from an Italian harpsichord: the jackslides are not the usual Italian boxslides, but separate upper and lower guides, a feature encountered in all north-European harpsichords and virginals (but not bentside spinets). It is equipped with two sets of unison strings, but has three rows of jacks, one of which plucks close to the nut, producing a penetrating, nasal timbre. A mechanism controlled by one of the stops projecting through the cheek probably disengaged this row of jacks when the other row (plucking the same string) was engaged. This contrast between the nasal, close-plucking stop and the normal register is a feature which could well be called German: it certainly does not occur in Italian harpsichords, with which the Müller shares some superficial similarities, but does occur in several 17th-century harpsichords from the German-speaking area. In addition, the instrument has a transposing keyboard which can be shifted to permit changing the sounding pitch of the instrument by a whole tone. Finally, the soundboard extends in a single piece across the gaps for the jacks so as to cover the wrest plank, a feature also encountered in other widely scattered instruments of the 16th and 17th centuries, including clavicytheria as well as harpsichords.

The bridge of the Müller harpsichord is not now in its original position (see fig.6). As it stands it was probably designed, in conjunction with a previous keyboard alteration, to give scalings suitable for brass strings at normal pitch. In its original position, and with the original keyboard capable of transposing by a tone, the bridge gives scalings suitable for iron stringing at a pitch an octave or a 9th above normal pitch depending on the position of the movable keyboard. Thus in addition to being the earliest surviving dated north-European harpsichord, it is one of the few examples from any period of an octave harpsichord.

The conceptions of instrument design embodied in the Müller harpsichord may have been carried to Flanders and, significantly, the earliest surviving Antwerp virginal was made in 1548 by Joes Karest, an emigrant who brought with him the Germanic tradition from Cologne. Like the Müller harpsichord, Karest's virginal has the look of a weightier version of an Italian design (in this instance the hexagonal virginal): although the case is thin, the mouldings are far heavier and the outline is less graceful than on a typical Italian instrument (see fig.26, p.114). In Flanders, virginals of this type began to be superseded in the late 1560s by instruments with thick cases, and it is possible that thick-cased harpsichords also began to be built then. The only evidence for this is the single-manual harpsichord that forms part of a claviorgan built in London by Lodewijk Theewes, an expatriate Fleming, in 1579. Although this highly important instrument preserves a number of features found in the 1537 Müller harpsichord (most notably the soundboard extending continuously to cover the wrest plank), its case is thicker and its scaling quite long (approximately 35.6 cm for *c″*): both these features correspond to later Flemish practice. The longer scaling produces a relatively shallow inward curve in the bentside and necessitates an appreciable shortening of scaling in the bass and a change from iron to brass wire. Whereas an Italian harpsichord would be likely to have string lengths that doubled for each octave throughout five sixths of the instrument's range, those of the Theewes harpsichord double down to only about *c′*, after which the bridge proceeds in a straight line instead of continuing to curve away from the spine. The bass section of the bridge on this instrument is not original, but on later Flemish instruments this section curves towards the spine at the end, instead of having a short section mitred on to it at an angle. The bracing system employed in the Theewes instrument is the first of those discussed earlier: a series of flat rectangular braces set vertically on the bottom of the instrument, and rising to the level of the underside of the case liners, run transversely from the bentside to the spine. This system, with minor variations, is found both in later English harpsichords and in German instruments of the Hamburg school.

The Theewes instrument is unusual for its early date in having three registers, two 8′ and one 4′, where one of the 8′ registers was probably used only as an arpichordum stop with brays, as on a Renaissance harp, on the 8′ bridge producing a buzzing sound on this register. As on most Italian instruments with a 4′ stop, all the tuning pins are placed at the front edge of the wrest plank, with the 4′ strings passing through holes drilled in the 8′ nut. Although this important instrument is essentially Flemish, it should be noted that it has a number of features that are probably characteristically English, including the use of oak for the case and the decoration of the inside of the case with embossed paper (both features can be seen in an English depiction of a virginal dating from 1591 and in surviving 17th-century English virginals). Moreover, the chromatic bass octave from C may also be a characteristically English feature; it is not normally found on continental instruments at this date (the 1550 Karest virginal and the Müller harpsichord of 1537, both in the Museo degli Strumenti Musicali, Rome, and both with a bass compass from *C* but lacking *C*♯ are either exceptions, or a reflection of an earlier practice), but it, too, can be seen in the painting of 1591 cited above. (The low *E*♭ appears in English keyboard music of the 16th century; the low *C*♯ may have been tuned to *A*′, although this note is not called for before Bull, Gibbons and Tomkins.)

The next surviving harpsichord of north-European origin is by Hans Moermans the elder of Antwerp, dated 1584. Much concerning the original state of this instrument must remain conjectural, but it now possesses many of the attributes of the developed Flemish types associated with the Ruckers family: a comparatively thick case made of softwood; a long scaling (35.6 cm for *c*″); soundboard barring consisting of a massive curving 4′ hitch-pin rail between the 8′ and 4′ bridges, with a diagonal cut-off bar to the left of the 4′ bridge and a group of transverse ribs stiffening the triangular areas between the cut-off bar and the spine; and bracing provided by deep transverse pieces attached to the bottom, supplemented by flat pieces nailed to the underside of the case liners and running obliquely from the bentside to the spine (the Moermans harpsichord, however, has only two braces of each type in addition to the divided belly rail, whereas later Flemish instruments have three). Unlike the wrest plank of the Theewes instrument, which is hollowed out underneath the nuts so that they stand on a freely vibrating soundboard area, and in which all the tuning pins are placed near the front, the wrest plank of the Moermans is solid and the 4′ tuning pins are placed between the 8′ nut and the 4′ nut, so that the 8′ nut does not have to be drilled to permit passage of the 4′ strings. The single row of 8′ jacks is placed in front of the row of 4′ jacks and plucks to the right, while the 4′ jacks in the back row pluck to the left.

3. *c*1590 TO *c*1700

(i) Flanders

(*a*) *The Ruckers family*. The developed Flemish harpsichord of the late 16th and the 17th centuries is inevitably associated with the work of the Ruckers family, a dynasty that dominated Antwerp harpsichord building for over a century, and whose instruments continued in use (sometimes radically rebuilt) throughout Europe as long as harpsichords were played.

The work of the Ruckers family is spread over four generations and almost 130 years, from 1579 to 1706. Hans Ruckers, the founder of the harpsichord-making business, was born in Mechelen between about 1540 and 1550; he died in Antwerp in 1598. His sons Joannes and Andreas became harpsichord builders and it is assumed that they were taught by their father. Joannes Couchet, the son of Hans's daughter Catharina, later continued Joannes Ruckers's business. By far the largest number of surviving Ruckers instruments are the work of Hans's sons: more than 35 by Joannes and some 35, dated between 1608 and 1644, made or signed by Andreas.

Hans Ruckers became a member of the Guild of St Luke, the Antwerp guild of artists, in 1579. In 1584 he rented a house in the Jodenstraat, Antwerp (Rubens, who painted the inside of the lids of some Ruckers instruments, was later to live nearby), and in 1597 he bought a property in the same street. Hans's work resembles that of older builders such as Hans Bos and Marten van der Biest, and in some respects that of Johannes (Hans) Grauwels, but it is not known whether he learnt his craft with any of these makers. (Links with van der Biest are known: he was a witness at Ruckers's wedding in 1575 and he was joined in Amsterdam by a man from Ruckers's workshop after 1585.) The few surviving instruments by Hans Ruckers are mostly virginals from the 1580s and 1590s. One combined single-manual harpsichord and virginal has survived; no double-manual harpsichord by him is known to exist. He was also an organ builder, but no example of this aspect of his work is known.

On the death of Hans (1598), his sons Joannes (baptized 1578) and Andreas (baptized 1579) became partners in the family business, but in 1608 Joannes bought out Andreas to become sole owner. The ledgers of the Guild of St Luke record the entry in 1611 of 'Hans Ruckers, sone, claversigmaker', evidently Joannes. Andreas does not appear in the records of the guild, but Jan Moretus, dean of the guild, mentioned him as a member in 1616–17 and in 1619 the guild ordered a harpsichord from him; he died between June 1651 and March 1653. He probably taught his only son, also Andreas (baptized 1607); at least seven instruments by the younger Andreas survive, from the 1640s and early 1650s (he died *c*1655).

Hans's grandson Joannes (Jan) Couchet (1615–55) was apprenticed to Joannes Ruckers in about 1627, worked with him until Ruckers's death (1642), and then took over the workshop and became a master builder. Although the instruments he built were essentially the same as those of Ruckers, Couchet began to introduce a number of innovations and to extend the compass and registrational possibilities. Only six of his instruments are known to exist. Three of Joannes Couchet's sons became harpsichord builders: Josephus Joannes (1652–1706), the most important, joined the Guild of St Luke in 1666 or 1667; Petrus Joannes and Maria Abraham also became members of the guild, but most of the existing instruments with a Couchet signature that are not by the elder Joannes Couchet seem to be by Josephus Joannes.

The array of models of virginals and harpsichords built by the Ruckers family is surprisingly large and varied. The virginals (see below and Chapter Two, §1) were built in a number of different sizes, each size designed to be tuned to one of a series of specific pitches covering a range of an octave plus a tone. The harpsichords were built with either one or two manuals and were of different sizes and pitches. Both single- and double-manual harpsichords were also built with an octave virginal combined in the same rectangular case structure.

The most common type of single-manual harpsichord was approximately 181 cm long and 71 cm wide, with one 8′ and one 4′ register, and with a split buff stop for the 8′ consisting of leather pads carried on sliding battens. The range of these instruments was almost invariably four octaves, *C/E* to *c‴*, although a few surviving examples originally had chromatic basses replacing the short octave, and sometimes extended to *d‴* in the treble. By the mid-17th century the Couchets regularly made single-manual harpsichords with a chromatic bass octave beginning at *C* and even with a keyboard extending chromatically down to *F′*, either of these sometimes with a 2 × 8′ disposition. Although no surviving instrument shows evidence of this disposition, documents show that the Couchets sometimes gave their instruments the more modern 2 × 8′, 1 × 4′ disposition. A late instrument, probably by Josephus Joannes Couchet (in the Nydahl Collection, Stockholm) was originally about 263 cm long and 92 cm wide and had a compass which

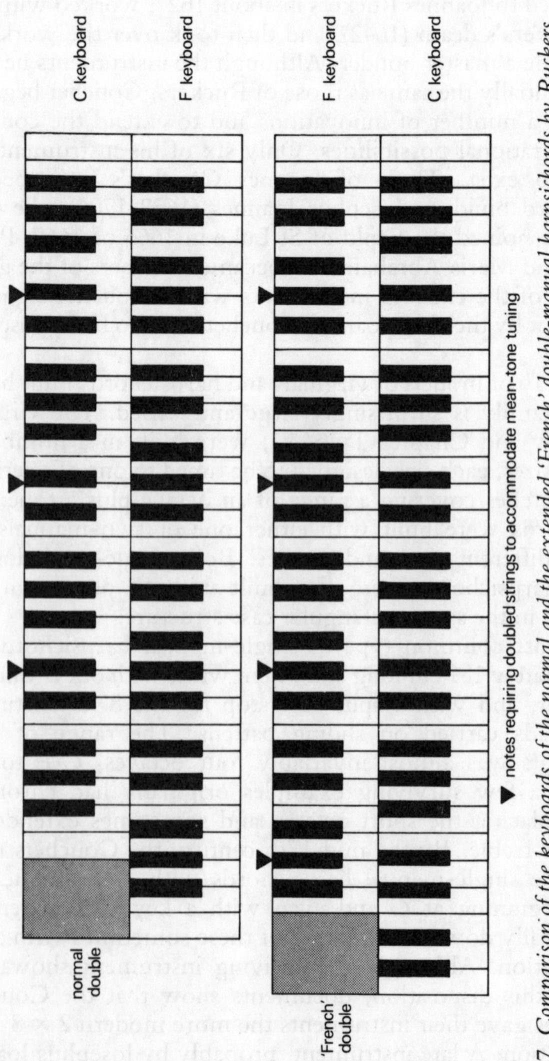

7. Comparison of the keyboards of the normal and the extended 'French' double-manual harpsichords by Ruckers

extended from *F′* to *e‴*, only one note short of the five-octave compass common by the mid-18th century.

A short, single-manual harpsichord (built by the elder Andreas Ruckers in 1627, now in the Gemeentemuseum, The Hague) is normal in having a compass of *C/E* to *c‴* and in having originally had a disposition with a single unison and a single octave choir of strings, but it is only 123 cm long. The short length derives from the fact that the strings throughout the entire compass are only two thirds of the length of the corresponding strings in a normal single-manual harpsichord. It is thus clear that this instrument was designed to be tuned a 5th higher than the more usual, longer, single-manual harpsichords. There is also documentary evidence and at least two surviving harpsichords (in the Metropolitan Museum of Art, New York, and the Nydahl Collection, Stockholm) to show that the Couchets built single-manual harpsichords designed to be tuned a major 2nd higher than the normal single-manual harpsichords.

Two-manual instruments may have been built in the Ruckers workshops as early as the 1590s. They had only two sets of strings, like the typical single-manual instrument, and only one of the two keyboards could be used at a time. The most common type of double-manual harpsichord (see fig.7) was about 224 cm long and 79 cm wide. The lower keyboard had 50 keys and a range of *C/E* to *f‴*, the upper had 45 keys and the smaller range of *C/E* to *c‴*. The pitch of the upper manual was the same as that of the usual type of single-manual harpsichord, and was about a semitone flat of modern concert pitch. (This pitch will subsequently be referred to as 'reference pitch' and denoted by 'R'.)

Slightly larger, extended-compass double harpsichords with a chromatic lower-manual keyboard *G′* to *c‴* and an *F* to *f‴* chromatic upper manual were also made. These instruments were of about the same length as the normal doubles, but were about 5 cm wider because of the addition of the chromatic notes in the bass. The pitch relationship between the keyboards of these instruments is the reverse of that of the normal Ruckers double (i.e. with the lower manual at reference pitch and the upper manual pitched a 4th below; see fig.7). This type of instrument seems to have been made specially for customers in France; the lower-manual compass extends below C and suits the music of Chambonnières and Louis Couperin, and that of the upper manual duplicates early French organ pitch and compass.

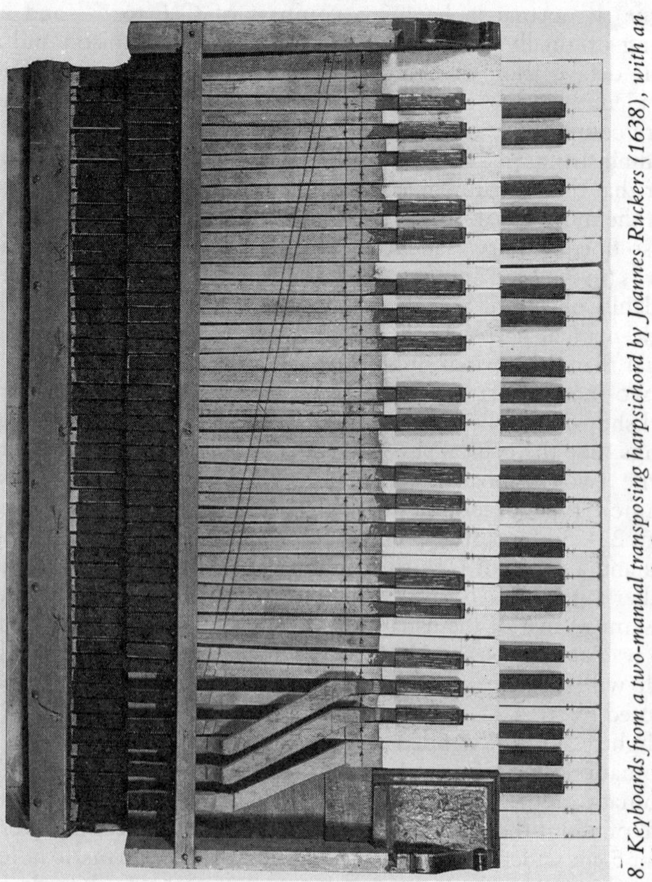

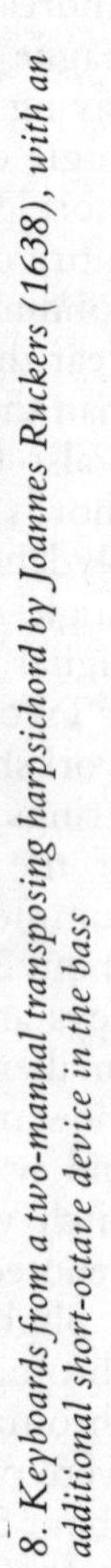

8. Keyboards from a two-manual transposing harpsichord by Joannes Ruckers (1638), with an additional short-octave device in the bass

The double-manual harpsichords were unlike most two-manual instruments in that the two keyboards were not aligned, but were positioned to sound a 4th apart. The *c'''* key of one keyboard was aligned with the *f'''* key of the other, and a block the width of three naturals filled in the space to the left of the lowest key of the upper manual. As a special refinement, extra strings were provided so that the *e♭* keys of the C keyboard would not be obliged to sound *d♯* corresponding to the *g♯* on the F keyboard. This required three doubled sets of strings for the normal double-manual harpsichord, and four doubled sets for the 'French' double (see fig.7). Because of these extra strings, the keyboards of such instruments could not use any rows of jacks in common; they were therefore completely uncoupled and had four rows of jacks for their two sets of strings. When one manual was being used, the jacks of the other manual were disengaged. Thus, a Ruckers double-manual harpsichord served as two instruments in one, playable at either of two pitches a 4th apart.

A single example of another unusual type of double-manual harpsichord exists (a much-altered harpsichord of 1612 by Joannes Ruckers, the property of HM the Queen; in Fenton House, Hampstead, London). This harpsichord was originally also about 224 cm long but only about 73.5 cm wide. One manual originally had a compass of *C/E* to *d'''* and was at a pitch of R – 5 (i.e. a 5th below reference pitch). The original pitch of the other manual is not known, but one highly likely possibility is that it was at R – 4, a tone above the other manual. This instrument also appears to have had only three registers, the central one probably a dogleg 8′ shared between the two keyboards, with the two outer 4′ rows playing one on each manual.

The Ruckers compound instruments combined a single- or double-manual harpsichord with a small virginal filling the space normally left outside the bentside of the harpsichord (see fig.9, p.38). In the combined single-manual plus virginal instruments, the virginal is at a pitch an octave above the harpsichord and therefore at R + 8. There is only one extant example of a Ruckers combined double-manual harpsichord and virginal. The keyboards in this instrument were originally of the type illustrated in the upper part of fig.7 (i.e. the normal double), and the pitch of the virginal was an octave above the lower manual. Since the lower manual is at R – 4, the virginal is

TABLE 1

Pitch scheme of the instruments built by the Ruckers family

pitch	pitch level	description
R + 9		2½-*voet* virginals
R + 8		3-*voet* or normal 'child' virginals
R + 5		4-*voet* virginals and 4-*voet* harpsichords
R + 4		4½-*voet* virginals
R + 2		5-*voet* virginals and some Couchet singles
R		6-*voet* virginals, normal single-manual harpsichords, one manual of the normal or extended-compass double-manual harpsichords
R – 4		one manual of the normal or extended-compass double-manual harpsichords
R – 5		one manual of Joannes Ruckers harpsichord of 1612

an octave higher than this at R + 5, and so is at quint pitch relative to reference pitch.

In all, the harpsichords and the combined harpsichords and virginals so far discussed represent six different pitches, namely: R – 5, R – 4, R, R + 2, R + 5 and R + 8. The Ruckers also made virginals at a number of different pitches, some of which were the same as these, some at the additional pitches of R + 4 and R + 9 (see Chapter Two, §1). The different types of virginals were specified in terms of their length in units of the Flemish foot or 'voet', which was subdivided into inches or 'duimen' (the *voet* had 11 *duimen*, and each *duim* was about 2.58 cm long; the Flemish foot is thus about 28.4 cm as opposed to the 30.5 cm of the English foot). Each type was at a different pitch. If the different pitches of the harpsichords and of the virginals are combined, instruments at altogether eight different pitches seem to group naturally into two 'consorts' a tone apart, with each 'consort' containing instruments separated in pitch from one another by 4ths, 5ths and octaves as shown in Table 1 (p.36). It seems unlikely, however, that these instruments were ever played together literally in consort. The problem of how these numerous pitch standards fitted into the daily musical life of the 16th and 17th centuries is as yet unsolved.

The construction of a Flemish harpsichord involved assembling the case sides (including the bentside), wrest plank and internal braces before installing the soundboard or the bottom. It was, therefore, a rather more complex task than the construction of an Italian instrument, which could be assembled by successively adding parts to the bottom, and in which the bentside was thin enough to be simply flexed and glued to the edge of the bottom; triangular supporting blocks (knees) were then attached to it. Instead, the parts of the Flemish case were held together by dowels and rabbets (or mortises), which helped to maintain everything in alignment during assembly. With the case and framing completed, the soundboard, already fitted with its bridges and ribs, was installed, and only then was the bottom attached to the lower braces and the lower edge of the case sides.

The natural keys of Ruckers harpsichords are covered with bone and the sharps are made of bog oak. The fronts of the natural keys are usually decorated with a punched paper design glued on to a layer of coloured parchment. At the back of the keyboard there is a slotted rack similar to that found in an Italian harpsichord. However, instead of a slip of hardwood to fit into

9. Single-manual Ruckers harpsichord with octave virginal in the bentside (date unknown): the rose types are those used by Joannes Ruckers after c1611 (see fig.11c–d, p.40)

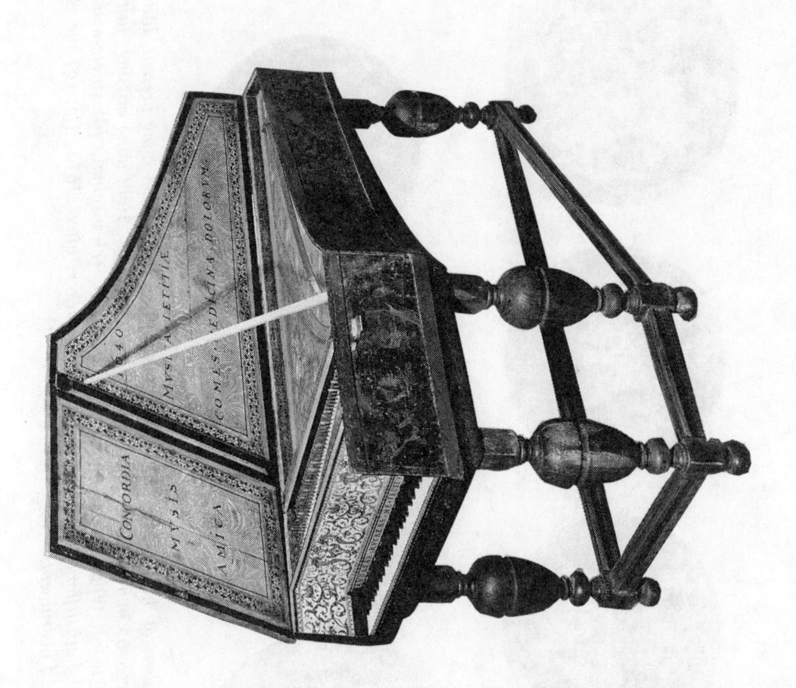

10. Harpsichord by the elder Andreas Ruckers, Antwerp, 1640

11. Soundboard roses used by members of the Ruckers family. (a) Hans Ruckers: angel's wing present, initials have rather flattened surface, diameter 65 mm; (b)–(e) Joannes Ruckers: (b) angel's right wing missing, initials have rounded upper surface, used in all types of instruments, c1596–1616; (c) virginal, (d) single-manual harpsichord and (e) double-manual harpsichord roses used by Joannes Ruckers after c1611; (f), (g) the elder Andreas Ruckers: used in all types of instruments

the appropriate slot in the rack, a Flemish keyboard has a metal pin driven into the end of the key-lever, and the rack is topped with a padded overrail that limits the upward motion of the keys. This system is also used in the lower manual of two-manual instruments; however, there is no space for a rack behind the keys of the upper manual of a two-manual instrument, and the tails of the upper-manual keys are therefore guided by vertical wires rising between the keys at the back of the plank on which the upper-manual balance rail is mounted.

The original decoration of Ruckers instruments was rather elaborate. Block-printed paper patterns (with motifs taken from Renaissance pattern books) were placed inside the key-well (above the keys) and above the soundboard around the inside of the case. These patterned papers were also sometimes used inside the lid in conjunction with a repeating wood-grained paper on which Latin mottoes were printed (see fig. 10); sometimes the insides of the lids were beautifully painted by contemporary artists such as Rubens, Jan Breughel and Van Balen. The outsides of the instruments were painted with an imitation of marble or of huge precious stones held in place by an ornamental iron strapwork. The soundboards were embellished with gouache paintings of flowers, birds, scampi, insects, snails, fruit and the like. The date was also painted somewhere on the soundboard or wrest plank. Decorative gilt roses placed in the soundboards incorporate the initials and trade mark of the builder, and are surrounded by a wreath or spray of flowers painted on the soundboard. All the roses of the Ruckers family represent an angel playing a harp, with the initials of the builder on either side of the angel. The exact posture of the angel and the layout and modelling of the rose varies from one member of the family to another and serves as one of the methods of determining the maker of the instrument (see fig. 11).

Only a few surviving Ruckers and Couchet harpsichords retain their original stands; contemporary paintings showing instruments of this kind reveal two common designs, either framed structures with thick, turned legs (see fig. 12, p. 43), or complex affairs with heavy, pierced-fretwork ends connected by arcades supported by numerous, turned balusters.

Ruckers' practice was to write a number on many of the action parts and on the case of each instrument as it was being made. The virginals were marked with the length of the instrument in Flemish feet. The harpsichords, both single- and double-manual, were marked with the letters 'St' for 'Staartstuk' (literally 'tailpiece') or harpsichord. Underneath this mark the serial number was written, a separate serial being used for each type of instrument. The existence of these numbers has made it possible for some undated instruments to be assigned an approximate date and for the rate of production to be estimated. The elder Andreas Ruckers, for example, made about 35 to 40 instruments a year when his workshop was in active production.

The tone of a two-register Ruckers harpsichord differs appreciably from that of an Italian instrument of the time, in having a more sustained brilliance and a somewhat less pronounced attack. The differentiation in timbre produced by the gradual change in plucking-point from a third of the string length in the extreme treble to about a tenth in the bass is adequate for distinguishing contrapuntal lines, but not so pronounced as to prevent the projection of a homogeneous sound in homophonic contexts. The 4′ register has a pleasant sound in its own right and is usable as a solo stop; the fact that the 8′ was capable of being disengaged means that it was almost certainly used in this way. The 4′ when combined with the 8′ lends a marked brilliance and carrying power to the ensemble. A buff stop can be engaged, which damps the higher overtones of the 8′ strings, producing a muted pizzicato effect. This buff stop was normally split between *c*′ and *c*♯′, enabling either the treble or the bass to be damped and contrasted with the sound of the undamped strings of the other half of the register. Registration was changed by reaching round the instrument and pushing or pulling extensions of the jackslides that passed through the treble cheekpiece, thereby moving the jackslide to the left or to the right to engage or disengage the register. Thus, the player could not change registers except during a pause between movements or individual pieces.

The importance of Ruckers instruments lies in their remarkable sound, which is the result of their extremely sophisticated design. The lengths, gauges and materials of the strings were chosen with great care. Soundboards and bridges were made of good materials and were accurately tapered to give the right thickness and stiffness in each part of the range. Also, the area of radiating soundboard was contrived to give an even balance between the bass, tenor and treble parts of the compass. The resulting sound is rich and resonant without any part of the register dominating another.

12. Flemish harpsichord painted in imitation of marble and decorated inside with block-printed paper: painting, 'The Music Master' (c1660), by Jan Steen

Ruckers instruments were justly famous in their own day, and their sound became an ideal during the 17th and 18th centuries in almost all of northern Europe. They were often altered and extended to suit later keyboard literature, sometimes by simple, even makeshift, alterations and sometimes by an elaborate rebuilding process involving the replacement of all the action parts and the extension and redecoration of the case. This process was commonly applied to double-manual harpsichords, the new keyboards being aligned to allow simultaneous use of contrasting registers. In France the process was known as *ravalement*. By leaving the original soundboard almost unaltered, the beauty of the sound could be preserved. In late 17th- and in 18th-century Europe, Ruckers and Couchet instruments were more highly valued than those of any other makers. Counterfeits were made with the decoration and appearance of genuine, rebuilt instruments, and existing instruments of suitable kinds were modified, given a fake signature and rose, and sold at an inflated price.

Ruckers instruments are important not only for their own beauty, but also because of their historical position as models for the later schools of harpsichord building. By the middle of the 18th century, the constructional methods of the indigenous schools of England, France, Germany, Flanders and the Scandinavian countries were securely based on the principles perfected by the Ruckers family. Soundboard design, action and stringing all reflect Ruckers practice, and the timbre is clearly reminiscent of Ruckers, even though characteristic also of the musical taste of the period and region.

There are now a number of well-restored Ruckers instruments, some in almost original condition, which can be heard in public concerts and on recordings. These instruments are extremely valuable as examples showing how they may once have sounded. However, restoration is not synonymous with preservation, as it nearly always involves loss as well as gain. The realization is thus growing that certain instruments should be left unrestored, in order that their extant original features may remain intact. (For a catalogue of extant Ruckers instruments see pp.48–56 below.)

(b) Non-Ruckers instruments. The scanty evidence available for harpsichord building before Hans Ruckers's admission to the Guild of St Luke in 1579 suggests that, as with the virginal, the basic characteristics of the single-manual harpsichord were established by the harpsichord makers of the preceding generation, one of whom would have trained Ruckers. In this connection it should be borne in mind that Lodewijk Theewes, who made the claviorgan of 1579 described above (see p.28), had emigrated to England at least seven years before he made that instrument, and that he had been admitted to the Flemish guild in 1561. The two-manual harpsichord appears to have been invented by a Flemish builder of Hans Ruckers's generation. Presumably the idea was suggested by the multi-manual organs of the time, with which Hans Ruckers would have been familiar in his capacity as an organ tuner and builder, but no double-manual harpsichord by him is known.

There are no surviving dated Flemish harpsichords from before 1584. The characteristics of the Flemish harpsichords of this period are therefore not certain, and it has to be assumed that they were similar in style to the extant Flemish virginals of this period. The two earliest Flemish virginals are dated 1548 and 1550 (in the Brussels Museum of Musical Instruments (see fig.26, p.114), and the Museo degli Strumenti Musicali, Rome, respectively) and both are by Joes Karest. Karest's name appears only once in the registers of the Antwerp Guild of St Luke, in 1523, when he was received as a master journeyman. Karest and his brother Goosen were among the other harpsichord and virginal builders who petitioned in 1557 for special recognition within the guild. Both of Karest's instruments are signed 'IOES KAREST DE COLONIA', indicating that he came from Cologne in Germany. The constructional details of these instruments are discussed below (see Chapter Two, p.115), but it should be noted here that they bear a very strong resemblance to the harpsichord by Hans Müller discussed above (see pp.25–8). This could suggest that the tradition of building harpsichords and virginals came to Antwerp and the Low Countries from the German-speaking part of Europe to the east of Flanders.

The next dated Flemish instrument, the so-called Duke of Cleves virginal of 1568 (in the Victoria and Albert Museum, London; see fig.28, p.120), is atypical in many respects, but is noteworthy in that it does not have the thin case sides with heavy applied mouldings of the Müller and Karest instruments. These, and the single-manual harpsichord built in 1579 by Lodewijk Theewes which also has the thick-cased construction typical of the later instruments, seem therefore to indicate that during the 1560s there was a transition in the style of building to the later tradition, found universally in all of the Flemish and Flemish-derived instruments, of using a heavy case with a decorative moulding cut in the top of the thick case side itself.

The two earliest signed and dated Flemish single-manual harpsichords, by Hans Moermans (1584) and Hans Ruckers (1594), originally possessed all of the features of case construction, soundboard barring, scalings, disposition, and action that are found in the later Flemish instruments. The surviving virginals from around 1580 by Hans Bos, Marten van der Biest, Johannes Grauwels and Hans Ruckers also show that by then virginal construction had settled into the pattern of later instruments. Harpsichord and virginal building was recognized as a profession within the Guild of St Luke after 1557, and it is clear that between 1560 and 1580 the characteristics of the later Flemish instruments were being established in the Antwerp workshops.

The earliest surviving double-manual Flemish harpsichord (an anonymous and undated instrument, definitely not by Ruckers, no.2934 in the Brussels Museum of Musical Instruments) was originally of the transposing type. The keyboards were laid out like those in the upper part of fig.7, but the compass of the upper manual was only *C*/*E* to *a''*, with a lower-manual compass of *C*/*E* to *d'''*. This compass suggests that the instrument was built about 1580. There were only three rows of jacks, probably with the middle row being a dogleg 8′ shared by both manuals, and the two outer rows were probably each a 4′, the near one operated by the upper manual and the far one by the lower manual. Because the same 8′ jacks were operated by both manuals, there was no possibility of using a system with doubled strings for the *e*♭/*g*♯ keys, and there is no evidence that the instrument ever had these. This would have meant that when transferring from one manual to the other it would have been necessary to retune the appropriate strings depending on whether a *g*♯ or an *e*♭ was required in the music. However, except for the smaller compass and the three-register disposition, this instrument is identical to the later Ruckers instruments except in small details such as construction marks.

The earliest dated Ruckers double-manual harpsichord, made by Joannes Ruckers and dated 1599 (in the Händelhaus, Halle), also originally had the *C*/*E* to *a''* upper-manual compass and *C*/*E* to *d'''* lower-manual compass, but it had four rows of jacks and doubled strings for the *e*♭/*g*♯ keys. Thus, except for the typically more limited compass it was a distinct improvement on the earlier, undated double in that no retuning of notes was required when changing manuals, and this type of construction set the pattern for double-manual harpsichords for at least the next 55 years.

Virginals and harpsichords built by other makers in Flanders in the period when the Ruckers were active are very similar to those of the Ruckers family. Indeed, most of the surviving harpsichords from this period not built by the Ruckers are anonymous instruments fraudulently attributed to them. The style of construction, the decoration, disposition, compass, scalings and (probably) sound are so similar in instruments built in the Ruckers tradition that it was an easy task for an 18th-century or modern forger to change the signature and rose of almost any virginal or harpsichord by a contemporary of the Ruckers and re-attribute it to one of that family without leaving himself open to suspicion. The only signed harpsichords that survive and are not by one of the Ruckers or the Couchets are the instruments by Hans Moermans the younger (1642), Simon Hagaerts (1632 and *c*1650), Gommarus van Eversbroeck (1659) and Joris Britsen (1675–81), and these exhibit the usual features of the Ruckers tradition. Hagaerts, like his Flemish contemporaries and the builders in most countries in Europe during the second half of the 17th century, used the 2 × 8′ disposition, which seems to have been very popular during this period.

CATALOGUE OF EXTANT RUCKERS INSTRUMENTS

Instruments are authenticated by comparing the dimensions, materials, decorations and construction marks with those of undoubted examples. Harpsichords with soundboards made from wood taken from what were probably Ruckers virginals are not included since the tapering of the resulting composite soundboard differs markedly from that of the Ruckers. Instruments dishonestly attributed to Ruckers by their makers, instruments altered and given fake inscriptions, roses, etc, and instruments referred to in the literature but otherwise lost, have not been included.

The virginals and single-manual harpsichords may be assumed originally to have had the compass *C/E* (short octave) to *c'''*, four octaves, unless otherwise specified. If no alteration is noted this is also the present compass; commas indicate missing accidentals (e.g. *G',A'* indicates *G♯'* missing). The double-manual harpsichords originally had their keyboards pitched a 4th apart (see fig.7). Unless otherwise specified these instruments had an upper-manual compass of *C/E* to *c'''* and a lower-manual compass of *C/E* to *f'''*. Double-manual instruments have been altered to align the pitch levels of their keyboards, unless otherwise noted.

The pitch at which the instruments sounded is determined by comparison of the length of their treble strings. The large virginals, 6 Flemish feet in length, most of the single-manual harpsichords, and the upper manual of the normal double-manual harpsichords all have, note for note, treble strings of the same length. The pitch of these instruments is here called 'reference pitch' and denoted by 'R', the pitches of the other instruments being referred to it. The 'mother' and 'child' instruments are all at reference pitch and an octave above respectively. (For illustrations of Ruckers 'mother and child' instruments see figs.9 and 30; see also fig.31.)

The virginals are referred to according to their length in Flemish feet. An instrument that is 6 Flemish feet long is called, for convenience, a 6′ instrument. The Flemish foot (*voet*) is about 28.4 cm instead of the 30.5 cm of the English foot. The mother virginals are 6′ instruments and the child virginals are of the spinett type. Rose types are shown in fig.11 (p.40).

Date, rose and instrument type	Compass	Remarks	Location
(1) HANS RUCKERS (*b c*1540–50; *d* 1598) Rose type *a*			
1581 HR Muselar mother and child virginals		Beautifully decorated with handpainted Renaissance motifs; lid painting	Metropolitan Museum of Art, New York
1583 HR 4′ spinett virginal a 5th above reference pitch	Present compass of *C/E*–*g*″, *a*″ is original	Unsigned except for rose; handpainted Renaissance decoration	F. Meyer, Paris
(*a*) 1591 HR Polygonal 6′ spinett virginal		Ruckers no.30; only extant 6-sided Ruckers virginal; bridge and jack-rail not original	Gruuthuuse Museum, Bruges
(*b*) [1591] HR Muselar mother and child virginals	Present compass of both is *C*–*c*‴ chromatic	Ruckers no.M/24; keys and nameboard of mother are modern; only lid painting dated	Skinner Collection, Yale U.
1594 HR 1-manual hpd with octave virginal in bentside	Present hpd *C*–*c*″ chromatic; virginal *C/E*–*g*″, *a*″ is original	Virginal has geometric rose	Kunstgewerbe-Museum, Schloss Köpenick, Berlin
(2) JOANNES RUCKERS (*b* 1578; *d* 1642) Rose type *b*			
1595 HR Child virginal	Present compass *C*–*d*‴ chromatic	Case and keys much altered; soundboard painting not original	Cincinnati Museum of Art
1598 HR 6′ spinett virginal	Present compass *G*′/*B*′–*c*‴	Ruckers no.6/61; papier-mâché rose; early example of paper decorations	Paris Conservatoire
1599 HR 2-manual hpd	Present compass *F*′–*f*‴	Much altered; earliest Ruckers 2-manual hpd	Händelhaus, Halle
1604 HR 5′ muselar virginal a tone above reference pitch		Ruckers no.5/34; signed 'Joannes et Andreas Rvckers me fecervnt'; papier-mâché rose	Brussels Museum no.2927

Date, rose and instrument type	Compass	Remarks	Location
1606 HR Spinett mother and child virginal		Ruckers nos.M/15 and k/15, date estimated from these; only extant spinett double virginal; paper decorations in excellent condition	Castello Sforzesco, Milan
1610 HR Muselar mother and child virginals	Both instruments now *C–f‴* chromatic	Numbered M/23 and k/23; case much altered and redecorated	Brussels Museum no.275
1611 HR 6′ muselar virginal		Ruckers no.6/16; AR rose not original; jack-rail and general style of instrument by Joannes Ruckers; has original stand	Vleeshuis Museum, Antwerp
(*a*) 1612 HR 2-manual hpd	Originally *C/E–d‴*; now *G′*, *A′–f‴*	Originally 2-manual hpd with one manual a 5th below reference pitch	Fenton House, London
(*b*) 1612 HR 2-manual hpd	*G′/B′–d‴* with split *B′/E♭* key	Ruckers no.St/34; originally normal Ruckers double; soundboard painting in fine condition	Musée d'Histoire Locale, Amiens
[1614] HR 6′ muselar virginal		Ruckers no.6/20; soundboard repainted and date obscured	Brussels Museum no.2930
1616 HR Extended-compass 2-manual hpd	Originally *F–f‴* and *G′–c‴*; now *F′–a″*, *b″*	Ruckers no.St/17; aligned extended 2-manual hpd; still only 2 choirs (1 × 8′, 1 × 4′)	M. Nirouet, Paris
Rose types: *c* virginal rose; *d* 1-manual hpd rose; *e* 2-manual hpd rose			
n.d. IR 1-manual hpd with octave virginal in bentside		Kbds and bridges not original; rose *c* in virginal; rose *d* in hpd	Berlin Museum no.2232

	Date	Maker	Instrument	Compass	Notes	Location
	1612	IR	2-manual hpd	Present compass F'–f'''	18th-century *ravalement* has left much of Ruckers's work intact	Paris Conservatoire
	1617	IR	2-manual hpd with ?virginal rose	Present compass G'–f'''	English *ravalement*, painted case and lid	R. Johnson, Los Angeles
(*a*)	1618	IR	Child virginal		Ruckers no.k/26; block-printed papers in excellent condition	Paris Conservatoire
(*b*)	1618	IR	2-manual hpd	Present compass G'/B'–d'''	Ruckers no.St/12; added keys taken from an Andreas Ruckers hpd; nut position not original	Schloss Cappenberg, Westphalia
(*c*)	1618	IR	2-manual hpd	Present compass C'–d''' chromatic	Soundboard, bridges, keys, decoration etc by J. C. Fleischer, 1724	Kulturhistoriska Museet, Lund
	1619	IR	2-manual hpd with quint virginal in bentside	Hpd G'/B' to c''; virginal C/E–c'''	Most of hpd, incl. soundboard, not original	Brussels Museum no.2935
	1620	IR	6′ muselar virginal	Present compass C–f''' chromatic	Ruckers no.6/3?; heavily restored	Museum of Fine Arts, Boston, Mass.
	1622	IR	6′ muselar virginal	Present compass C–f''' chromatic	Ruckers no.6/38; much restored; some original papers	Metropolitan Museum of Art, New York
	1623	IR	Muselar mother and child virginals		Ruckers nos.M/33 and k/33; date estimated from these	Württembergisches Landesmuseum, Stuttgart
	1624	IR	2-manual hpd	Present compass G', A'–d'''	Soundboard painting in exceptional condition; French keys and action	Musée des Unterlinden, Colmar
(*a*)	1627	IR	1-manual hpd	Present compass C–e''' chromatic	Ruckers no.St/54; widened case and keyframe; now 2 × 8′	Berlin Museum no.2227
(*b*)	1627	IR	2-manual hpd	Present compass G'–c'''	Alignment dated 1701; some original papers	P. de la Raudière, Château de Villebon, Eure-et-Loir
(*c*)	1627	IR	2-manual hpd, originally of extended compass	Present compass F'–f'''	*Petit ravelement* had G'–d''' aligned kbds by builder who altered 1632 IR	Private ownership, Switzerland

Date, rose and instrument type	Compass	Remarks	Location
(*a*) 1628 IR Mother muselar virginal	Present compass *C, D–f'''*	Numbered M/34; marquetry veneer on case a later addition	Brussels Museum no.2926
(*b*) 1628 IR 2-manual hpd, originally of extended compass	Present compass *G'–d'''*	Important hpd with decoration and compass contemporary with F. Couperin	Versailles Palace
1629 IR $4\frac{1}{2}'$ spinett virginal a 4th above reference pitch	*C/E–d'''* compass is original	Ruckers no.$4\frac{1}{2}$/11; has its original jacks	Brussels Museum no.2511
1632 IR 2-manual hpd, originally a normal 1-manual hpd	Present compass *F'–e'''*	Elaborate Louis XV decoration, *ravalement* by same builder as (*b*) 1628 IR	Musée d'Histoire et d'Archéologie, Neuchâtel
1636 IR 6' muselar virginal	Present compass *C–f'''* chromatic	Ruckers no.6/70; papers not original; bridges moved and repinned	Harvard U., Cambridge, Mass.
(*a*) 1637 IR 1-manual hpd	Present compass *A'–f'''* (originally *C–c'''* chromatic)	English *ravalement* and decoration; restored to playing condition	Russell Collection, U. of Edinburgh
(*b*) 1637 IR 2-manual hpd	Original (unaligned kbds)	Ruckers no.St/14; kbds aligned 18th century; restored to original state 1972	Museo dei Strumenti Musicali, Rome
(*a*) 1638 IR 6' muselar virginal	Present compass *C–d'''* chromatic	Ruckers no.6/68; has its original lid papers	Brussels Museum no.2933
(*b*) 1638 IR 2-manual hpd	Original (unaligned kbds)	Ruckers no.St/41; fine lid painting and paper decoration in good condition	Russell Collection, U. of Edinburgh
1639 IR 1-manual hpd	Original compass *C–d'''* chromatic; kbd missing	The only extant example of this unusual chromatic compass	Victoria and Albert Museum, London
(*a*) 1640 IR 5' muselar virginal a tone above reference pitch	Present compass *C–c'''* chromatic	Ruckers no.5/46; many original papers	Gemeentemuseum, The Hague

(*b*) 1640 IR 2-manual hpd	Present compass *G′/B′–c‴*	Ruckers no.St/14; fine lid painting, restored to playing condition	Erbdrostenhof, Münster
(*a*) 1642 IR 5′ muselar virginal a tone above reference pitch		Ruckers no.5/78; has IR rose of design similar to Joannes Couchet virginal rose	Musikmuseet, Stockholm
(*b*) 1642 IR 2-manual hpd	Present compass *G′/B′–d‴*	Ruckers no.St/24; fine lid painting	H. Gough, New York
(3) ANDREAS RUCKERS (i) (*b* 1579; *d* after 1645) Rose type *f* unless otherwise noted			
[1605] AR 1-manual hpd	Kbd missing	Ruckers no.St/2; date estimated from this	Vleeshuis Museum, Antwerp
1608 AR 2-manual hpd	Present compass *G′/B′–d″*	Ruckers no.St/19; converted to pf in 18th century; nut positions altered	Russell Collection, U. of Edinburgh
1609 AR 1-manual hpd	Present compass *C–d‴* chromatic	Soundboard painting in fine condition; English action and veneered case	P. Williams, Edinburgh
(*a*) 1610 AR $2\frac{1}{2}$′ spinett virginal a 9th above reference pitch		The only extant example of this type of virginal, sloping case-sides, fine lid painting	Private ownership, Australia
(*b*) 1610 AR $4\frac{1}{2}$′ muselar virginal a 4th above reference pitch		Ruckers no.4x/35; rose not original	Museum of Fine Arts no.295, Boston, Mass.
(*a*) 1613 AR 4′ spinett virginal a 5th above reference pitch		Ruckers no.4/11	Brussels Museum no.274
(*b*) 1613 AR 4′ spinett virginal a 5th above reference pitch		Ruckers no.4/40	Brussels Museum no.2928
(*c*) 1613 AR Child virginal	Present compass *C/E–d‴*	Kbds not original; decoration in style of an early work of (3) Andreas Ruckers (i)	Cincinnati Museum of Art

Date, rose and instrument type	*Compass*	*Remarks*	*Location*
1614 AR 2-manual hpd	Present compass *A′–e‴*	English *ravalement*, action and veneered case	L. Elmhirst, Dartington Hall, Devon
1615 AR 2-manual hpd	Keys missing; originally normal 2-manual hpd	Ruckers no.4; has its original paper decorations and unusual marbled exterior	Vleeshuis Museum, Antwerp
1617 AR 6′ spinett virginal	Present compass *C–f‴* chromatic	Ruckers no.6/23	Deutsches Museum, Munich
1618 AR 1-manual hpd		Ruckers no.St/1; has original kbds, scalings; 1 × 8′, 1 × 4′	Berlin Museum no.2224
(*a*) 1620 AR 4′ spinett virginal a 5th above reference pitch		Ruckers no.4/69; has its original printed papers and key arcades	Smithsonian Institution, Washington, DC
(*b*) 1620 AR 6′ muselar virginal	Present compass *C–c‴* chromatic	Ruckers no.6/27	Brussels Museum no.1597
(*c*) 1620 AR 2-manual hpd	Present compass *G′/B′–f‴*	Ruckers no.St/68; has its original stand	Berlin Museum no.2230
(*d*) 1620 AR 6′ muselar virginal	Present compass *C–g‴* chromatic	Much altered; bridges not original; many original papers	Mme Labeyrie, Gif-sur-Yvette
1621 AR 2-manual hpd, originally 1-manual	Present compass *F–f‴*	French *ravalement* using parts of original soundboard and case	Private ownership, France
1623 AR 2-manual hpd	Present compass *F′*, *G′–f‴*	English *ravalement* of normal double with case widened in treble; veneered outer case	Private ownership, England
1624 AR 1-manual hpd	Present compass *F′–f‴*	Originally a normal 2-manual hpd of normal specification	Gruuthuuse Museum, Bruges
[1626] AR Child virginal	Present compass *C/D–c‴*	Ruckers no.k/36; date estimated from this	Sterckshof Museum, Deurne, Belgium
1627 AR 1-manual hpd a 5th above reference pitch		Ruckers no.4/St; decorations in excellent condition; original stand	Gemeentemuseum, The Hague

	1628 AR 2-manual hpd	Present compass *A′–d‴*	Formerly attrib. (2) Joannes Ruckers but in style of (3) Andreas Ruckers (i)	Private ownership, Germany or Switzerland
	1632 AR 4′ spinett virginal a 5th above reference pitch	Present compass *C–f‴* chromatic	Ruckers no.4/38; bridge position and much else altered	Brussels Museum no.1593
(*a*)	1633 AR 6′ muselar virginal	Present compass *C–f‴* chromatic	Ruckers no.6/70; added keys from another 6′ virginal	Brussels Museum no.4600
(*b*)	1633 AR 1-manual hpd	Present compass *C/D–d‴*	Ruckers no.St/41; originally 2-manual hpd	Karl-Marx-Universität, Leipzig
	1634 AR Child virginal	Present compass *C–a″* chromatic		Paris Conservatoire
	1635 AR 1-manual hpd	Present compass *C*, *D–c‴*	Much of soundboard missing; lid papers and some case papers extant	M. Thomas, London
	1636 AR 2-manual hpd	Present compass *F′–f‴*	Originally 1-manual hpd *C–c‴*; much altered; *ravalement* probably by Hemsch	M. Thomas, London
	1637 AR 1-manual hpd		Numbered with both St/23 and St/24; original disposition and decoration; rose not original	Germanisches National-museum, Nuremberg
(*a*)	1639 AR Child virginal		Ruckers no.k/59; rose type *g*	Gemeentemuseum, The Hague
(*b*)	1639 AR 1-manual hpd	Former compass (as 18th-century double) *G′–d‴*	Converted to a double, 18th century, now back to a single; rose type *g*	Gemeentemuseum, The Hague
(*a*)	1640 AR 1-manual hpd	Present compass *C–d‴* chromatic	English kbd and action; original decorations; rose type *g*	Skinner Collection, Yale U.
(*b*)	1640 AR 2-manual hpd	Keys missing	Ruckers no.St/2; heavy gessoed decoration not original	Musée de Croix, Namur, Belgium

Date, rose and instrument type	Compass	Remarks	Location
(*a*) 1643 AR 5′ muselar virginal a tone above reference pitch	Present compass *C–c‴* chromatic	Ruckers no.5/37; rose missing, but decorations in the style of (3) Andreas Ruckers (i)	Gemeentemuseum, The Hague
(*b*) 1643 AR 2-manual hpd	Present compass *G′–e‴*	Conservative French *ravalement* of a normal double; rose missing	Private ownership, USA
(*a*) 1644 AR 1-manual hpd	Present compass *C–c‴* chromatic	Ruckers no.St/16; original decorations; signed 'Andreas Rvckers den Ovden me fecit'	Vleeshuis Museum, Antwerp

(4) Andreas Ruckers (ii) (*b* 1607; *d c*1655)
Rose type *g*. The distinction between the instruments of (3) Andreas Ruckers (i) and (4) Andreas Ruckers (ii) is not clear. Instruments dated after 1637, when (4) Andreas Ruckers (ii) entered the Guild of St Luke, have rose type *g* and many show positive signs of being by the younger Andreas.

Date, rose and instrument type	Compass	Remarks	Location
(*b*) 1644 AR Muselar mother and child virginals	Both instruments now *c*, *D–c‴*	Ruckers no.M/28; child a modern replacement	Karl-Marx-Universität, Leipzig
(*a*) 1646 AR 1-manual hpd	Present compass *C–f‴* chromatic		Vleeshuis Museum, Antwerp
(*b*) 1646 AR 2-manual hpd	Present compass *F′–f‴*	French *ravalement* signed 'Refait par Pascal Taskin, Paris 1780' of extended-compass double hpd	Paris Conservatoire
1648 AR 1-manual hpd	Present compass *C–d‴* chromatic	Ruckers no.St/69; original decorations; English kbds	Musikhistorisk Museum, Copenhagen
(*a*) [1651] AR 1-manual hpd	Present compass *C–d‴* chromatic	Ruckers no.St/2; original decorations; date ?1641	P. Maxwell-Stuart, Traquair House, Peeblesshire
(*b*) 1651 AR 2-manual hpd	Present compass *G′*, *A′–f‴*	Originally 1-manual hpd; English *ravalement*	Victoria and Albert Museum, London
1654 AR 2-manual hpd	Present compass *G′/B′–f‴*	Originally 1-manual hpd	Germanisches Nationalmuseum, Nuremberg

(ii) France

Notwithstanding the great importance of Flemish harpsichords and the fact that they were shipped all over Europe and even to the Spanish colonies of the New World, harpsichords of a different kind were being made in England, France and Germany, and probably also in the Iberian peninsula. Only a relatively small number of these survive, and they seem more closely related to such instruments as the Müller harpsichord and the Karest virginal than to either Italian instruments or developed Flemish instruments of the Ruckers type.

Much still needs to be learnt about these largely vanished types, especially since so much important harpsichord music was composed for them rather than for the more numerous instruments of the mid- and late 18th century. Whereas Flemish harpsichords had exclusively iron scalings and most 17th-century Italian harpsichords had brass scales, in France both scalings were in use. Since brass-strung harpsichords tend to have a bolder sound with more attack (as a result of the string material) there are basic differences in the two types of instrument. Whether one of these types was more common in the early 17th century is hard to tell; in the latter half of the century iron scales predominate. Soundboards are often cross-barred and the cases have thicknesses intermediate between Flemish and Italian traditions, so that French instrument making at this period cannot be aligned exactly with its neighbours to the north or south. The few playable examples of these harpsichords are less rich and full than the Flemish or later French examples, but they are well-balanced in themselves and have a clarity of musical speech that is not available in a more resonant instrument. However, such generalizations are always dangerous since individual instruments can easily deliver a counter-argument. Some of these instruments (including the clavicytherium depicted by Praetorius), though by no means all, have round tails rather than oblique straight ones, so that their bentsides run from the cheekpiece to the spine in a continuous S-shaped curve.

The internal structure of all the surviving instruments of the intermediate type seems to lack the upper level of braces found in the instruments of the Ruckers type, and the sides are assembled around the bottom rather than the bottom being added to an otherwise completed structure. A number have only struts sloping downwards from the bentside liner, others have thin sides supported by triangular knees in the Italian manner, and still others have deep braces running obliquely across the case from the bentside to the spine and belly rail. The types of soundboard barring are at least as variable, with some examples barred exactly as in Ruckers harpsichords except for the use of a less massive 4′ hitch-pin rail, and others having transverse ribs extending right across the soundboard, occasionally but not invariably cut out under the bridges and occasionally even dividing the 4′ hitch-pin rail into separate segments.

Examples of only two representative species survive in any quantity: the well-known English virginals of the mid- and late 17th century, with their characteristically vaulted lids (seen also in pictures of French virginals in the 16th and 17th centuries); and French harpsichords of the second half of the 17th century, almost all of which (in contrast to the majority of Flemish, Italian and English instruments of the period) have two manuals. Whether this latter fact is an accident of preservation is not clear; single-manual instruments are mentioned in inventories of the period and only a single-manual instrument is depicted by Mersenne in his *Harmonie universelle* (1636–7), but it remains true that only doubles survive in any quantity from France. All these are of the contrasting type with aligned keyboards.

The cases of these French harpsichords are generally made of walnut, although some have a spine made from pine. A characteristic feature is the application of a decorative scroll-sawn piece to the inside of the spine and the cheekpiece at each end of the keyboard, a piece resembling those found in the Italian instruments with thick cases and cypress linings intended to counterfeit the appearance of a thin-cased instrument in its outer case. The keyboards themselves are particularly elegant, the natural keys covered with ebony and the sharps usually made from solid ivory blocks. Instead of carved or cut-out arcades on the fronts of the natural keys, a recessed trefoil pattern is formed into the front of the key by deep, slanting cuts with a round gouge, a method of decoration also seen in some German and Spanish keyboards.

As in a Ruckers instrument, the backs of the lower-manual keys are guided by metal pins fitting into the slots of a wooden rack (see fig.14, p.60). The keys of the upper manual are, however, guided in a way unique to French instruments, a way that continued in use throughout the 17th and 18th centuries.

13. Harpsichord by Jean-Antoine Vaudry, Paris, 1681

Instead of pins placed between adjacent keys, each key has its own pin which fits in a narrow longitudinal slot cut through the key-lever near the back. These pins are set in the back rail of a key-frame (resembling that which supports the lower manual) rather than in a plank like that of both Flemish and later English instruments.

Originally the range of these keyboards was almost invariably G'/B' to c''', occasionally with one or both of the two lowest sharps divided to permit sounding the missing accidentals. From

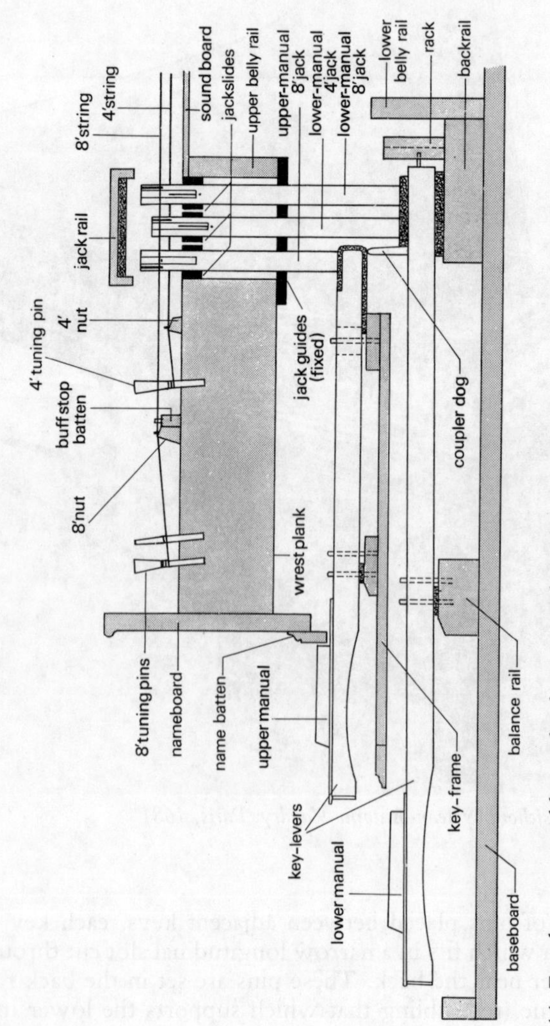

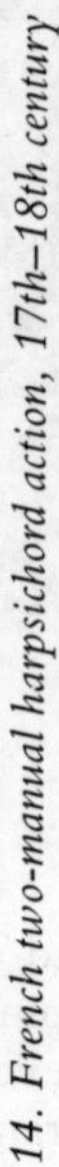
14. *French two-manual harpsichord action, 17th–18th century*

the 18th century this range was sometimes enlarged by sacrificing the endblocks on the key-frames and the scroll-sawn cheekpiece liners to make room for a chromatic bass (sometimes omitting the *G*♯′) and a treble expansion to *d*‴.

Some 17th-century French harpsichords do not appear to have been painted. Their graceful walnut cases, sometimes decorated with marquetry, were simply varnished, as were their rather elaborate stands with six or more spiral legs and framing or stretchers placed above ball feet.

Information about this highly important part of the history of the harpsichord can be gleaned from written sources, as well as from the French instruments and a handful of English and German examples. Praetorius's illustrations to vol.ii of *Syntagma musicum* (1618; published in 1620 as *Theatrum instrumentorum*) show a harpsichord of this type, which seems to have had two 8′ registers and one 4′, although Praetorius's text states that four registers, including one tuned a 5th above the 8′, were possible. Mersenne also mentioned the possibility of such a quint register, although his illustration shows only a single 8′ and 4′ with both rows of tuning pins set at the front edge of the wrest plank. Mersenne was concerned with the ways in which the harpsichord can produce changes in loudness and timbre, and stated that the instruments of his time could produce seven or eight different combinations of register. Seven combinations can be achieved on a three-register instrument; eight implies a buff stop used only with one of the registers when it was being played alone, or else a four-register instrument in which one register was tuned to a different pitch from the others in order to provide an equivalent to the Flemish transposing double (perhaps this was the actual function of the quint register); but Mersenne made no explicit statement on this matter.

Mersenne also mentioned instruments with two or even three keyboards. Most information on French two-manual instruments of the first half of the 17th century, however, must be derived from other sources, most notably a letter of 1648 from Pierre de la Barre, harpsichordist to Louis XIV (in the correspondence of Constantijn Huygens, ed. J. A. Worp, The Hague, 1911–17). In this letter La Barre distinguished between French and Flemish harpsichords on the basis of a recent invention that gave each keyboard of a French two-manual instrument its own set of unison strings, whereas on Flemish instruments the unison strings were shared by the two manuals. Although it is possible to interpret this as a reference to the Flemish transposing double with its single set of 8′ strings, it may refer to a dogleg jack that causes the upper-manual 8′ to sound from the lower manual as well whenever the upper-manual 8′ was engaged. If this hypothesis is correct, the invention to which La Barre referred was the manual coupler by means of which the player has the choice of playing the upper-manual 8′ register from the lower manual or not.

This simple and elegant device consists of equipping the lower-manual keys with small vertical pieces of wood that rise to the level of the underside of the upper-manual keys, and then allowing one of the two keyboards to be slid in and out of the instrument by rather less than a centimetre while the other is held in position. In one position, the vertical pieces of wood, called 'coupler dogs' (see fig.14), are located directly below the ends of the upper manual, so that when the keys of the lower manual are played the coupler dogs push upwards on the ends of the upper-manual keys, causing the upper-manual register to sound at the same time as the lower-manual registers. Thus, when the keyboards are in this position, the instrument functions much as if it had dogleg jacks. If, however, the upper keyboard is pulled outwards or the lower keyboard is pushed inwards, the coupler dogs are positioned beyond the ends of the upper-manual keys, so that the two keyboards are completely independent and may be used to produce simultaneous differences in timbre between the part played by the right hand and that played by the left, and the same note may even be doubled simultaneously by both hands. (This possibility was exploited in a typically French type of harpsichord piece termed the *pièce croisée*, the earliest extant examples of which were written by Louis Couperin, who died in 1661.)

The exact date when the manual coupler appeared is difficult to establish. La Barre's reference to a recent invention may well have to be taken with some scepticism, since Mersenne in 1636 had described 'little wedges of wood at the backs of the keys by means of which one can effect changes of registration', and it is difficult to imagine what else he might have been thinking of, even though the relevant passage seems concerned otherwise only with single-manual instruments. In the 17th century it seems that coupling was effected by pulling the lower manual outwards; by the 18th century, however, the lower manual was fixed and the upper manual was pushed inwards. However, some 17th-century French double harpsichords were originally disposed without a coupler, but rather with a dogleg front register and two registers on the lower manual. Most such instruments were later altered by adding a coupler and removing the dogleg from the upper-manual jacks. There is also evidence that the dogleg register or independent upper register was sometimes a 4′. A notable surviving example is the 1693 Nicolas Blanchet double harpsichord in a private collection in Paris (see §4(i) below). It has a coupler and two 8′ registers on the lower manual; all the upper-manual jacks are missing but they must have been of a 4′ register.

With the creation of the coupler, the resources of the classic harpsichord of the 18th century were all available, at least in theory. An instrument with two independent manuals served by three registers (2 × 8′, 1 × 4′) can play all of the great harpsichord repertory; and although instruments with greater resources certainly existed before the middle of the 18th century, they must be regarded as somewhat exceptional.

(iii) England

The small group of surviving English harpsichords of the intermediate type include an oak instrument of 1622, which appears to have had three 8′ registers of different timbres, and a round-tailed instrument made by Charles Haward in 1683 with two sets of 8′ strings and originally with a row of close-plucking lute-stop jacks to produce a different timbre from one of them, much as on the 1537 Müller harpsichord. The earliest known dated two-manual English instrument (apart from an example bearing a somewhat doubtful date of 1623) was made by Joseph Tisseran in 1710 and has three registers – a dogleg 8′, a lower-manual 8′ and a 4′; but the disposition that was later to become typical in England is found in an instrument made by Hermann Tabel in 1721 (see fig.15, p.64). It has three sets of strings (2 × 8′, 1 × 4′) and four rows of jacks: a dogleg 8′, the lower-manual 8′ and 4′ at the main gap at the back of the wrest plank, and a row of close-plucking lute-stop jacks set in a register that runs diagonally across the wrest plank near the 8′ nut and plucks the same strings as the dogleg jacks. The 8′ and 4′ scalings have been altered. A harpsichord made by Thomas Hitchcock in about 1725 (see fig.16, p.67) now has the same disposition as the Tabel instrument. However, the Hitchcock instrument has been modified: it appears to have had a manual coupler originally,

15. Harpsichord by Hermann Tabel, London, 1721

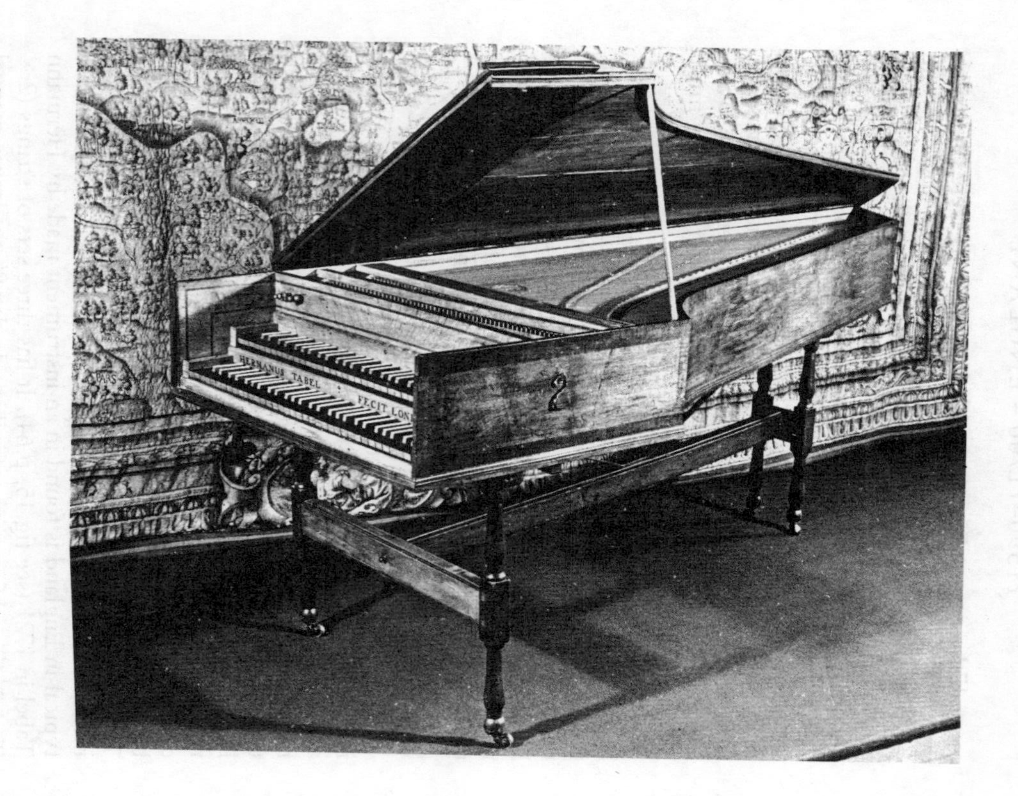

instead of the dogleg jacks, and the lute stop may have been added later.

Characteristically, the two rows of unison jacks are placed next to one another with the 4′ at the back rather than between the 8′ rows. This arrangement tends to reduce the contrast between the timbres of the two 8′ registers by reducing the difference in the points at which their strings are plucked. This, in turn, improves the blending of the two unison registers when they are both engaged; and since they can never be heard singly in rapid succession owing to the presence of the dogleg, which requires the upper-manual 8′ to be disengaged to hear the lower-manual 8′ by itself, the lack of contrast in timbre between the two registers cannot be considered a defect. (This may also be the reason why a number of single-manual instruments from countries where the 4′ jacks would be found between the 8′ rows on a double-manual also have their 8′ rows adjacent to one another.) On instruments with a manual coupler, the difference in timbre between the two unison stops may easily be exploited by the performer and is particularly advantageous in distinguishing between the two voices of *pièces croisées*. At the same time a pronounced difference in loudness between the two 8′ registers, which might be desirable in voicing either a single-manual instrument or one with a dogleg, cannot be permitted, as the performance of *pièces croisées* requires a similar degree of loudness on each manual.

On instruments like the Tabel, the close-plucking lute stop (available only on the upper manual) provides a contrasting unison register that may be used in dialogue with the lower-manual 8′, despite the presence of the dogleg unison register, since with the lute stop engaged the upper manual is not silenced when the dogleg register is disengaged. But, because the principal function of the second manual on a two-manual harpsichord was to permit contrasts in loudness, such an independent upper-manual stop was not a necessity on an instrument with dogleg jacks, and it did not appear on Flemish harpsichords until the 18th century (and then only on some of them).

An English development of little influence at the time but of considerable influence on harpsichords of the modern revival was the creation of a foot-operated mechanism for effecting changes of registration. According to Thomas Mace (*Musick's Monument*, 1676), the 'pedal' was a recent invention of John

Hasard (not Haward, as long supposed; see Mactaggart, 1987) of London. The instrument had four pedal-controlled registers (most probably two 8′, a lute stop and a 4′), and the example owned by Mace also had a hand-operated buff stop. Nothing further is heard of pedal-controlled registers until the third quarter of the 18th century, when their primary function was that of producing crescendo effects rather than of changing registration in a framework of terrace dynamics, and Hasard's idea was really taken up only when 'modern' harpsichords began to be made in the last years of the 19th century.

(iv) Italy

To organize the history of instruments into centuries is to risk introducing divisions where continuities should perhaps be stressed. However, in the case of Italian harpsichords there is some justification for making a division between the 16th and 17th centuries. The history of these harpsichords is not the 'virtually seamless continuum' from the 16th century to the 18th (Ripin, *Grove 6*, 'Harpsichord', p.234) which has been reported. An impression of uniformity has been conferred by the many alterations of 16th-century instruments.

It has been described above (see p.15) how many of the instruments with a 1 × 8′, 1 × 4′ disposition were altered to 2 × 8′. These alterations took place in the 17th century, although the change did not take place suddenly. An instrument made by Trasuntino in 1570 had already had its 4′ removed by the 1630s. Some other alterations can be dated to the second half of the 17th century. From the dated instruments the impression gained is that 1 × 8′, 1 × 4′ harpsichords were no longer built around 1600. The latest 1 × 8′, 1 × 4′ harpsichord of known date was made in 1585, now carrying the name 'Bertolotti' (although the maker was probably Bonafinis). In fact there are several instruments which might have been built around 1600, but a precise dating is not possible.

Most of these 1 × 8′, 1 × 4′ harpsichords had a *C*/*E* to *f*‴ keyboard. In many instances this was modified (probably at the time the 4′ was removed) to give a *G*′/*B*′ to *c*‴ keyboard. This was a relatively easy modification since both keyboards had 50 notes and the whole stringband did not have to be altered. As a result of this change, the new *c*″ key (of the *G*′/*B*′ to *c*‴ keyboard) then played the old *f*″ string. The harpsichord was restrung in brass wire (perhaps with a minor modification to the

16. Harpsichord by Thomas Hitchcock, London, c1725

nut position in order to adjust the string length) and therefore sounded about the same pitch as before. Although the keyboard was changed, it is not entirely clear whether this was done because the *C/E* to *f'''* compass was no longer satisfactory, or because the sound of a brass-strung harpsichord was preferred. On balance, it seems that the unsuitability of the *C/E* to *f'''* compass was the reason for this change since there were fewer new *C/E* to *f'''* harpsichords built after the middle of the 17th century. On the other hand, there are a few clues that some instruments may have been rebuilt because of the preference for the sound of brass wire.

Towards the end of the 17th century a range of *G'/A'* to *c'''* was widely used. Since smaller compasses were in use before this, the conclusion follows quite naturally that there was a gradual increase in the range of the music that the instruments were used for. Certainly, this is the trend for instruments in other countries. However, when the dates of the appearance of these compasses are considered, then some curious patterns emerge. The *C/E* to *f'''* compass is known as early as 1523 and was easily the most common compass in the 16th century (which is also found in the virginals of this period; i.e. it is a phenomenon of keyboard instruments and not just of harpsichords). However, in the 17th century it gives way to a smaller compass: *C/E* to *c'''*. It may not be true that the compass *G'*,*A'* to *c'''* was only a late 17th-century occurrence since a wide range was conceivably in the 1538 Dominicus Pisaurensis harpsichord (*G'*,*A'* to *a''*; see p.19 above). Thus an orderly progress from smaller compasses to larger ones cannot be found. Explaining how the *C/E* to *f'''* compass was used belongs to the realm of performing practice since very little music that reached above *c'''* was written in the 16th or 17th centuries in Italy. Furthermore, the *C/E* to *f'''* keyboards were not at a specially low pitch because of their compass. This means that the extension in the treble was, in effect, an extra register. It is known that organs were built at 4′ pitch at this time, as well as octave harpsichords and virginals. Thus the use of 4′ pitch in itself should not be surprising. To what extent compasses were made reaching to *f'''* in order to facilitate upward transpositions has not been determined.

Many points of detail concerning 16th- and 17th-century performing practice are not yet clear enough to illuminate the difficulties here. Concentration should therefore be on the general direction of development and it should be noted that this was the period in which the harpsichord took on an important role in providing a continuo line. Thus the changes seen in the harpsichord's compass in the 17th century were responses to the changing role of the instrument. There was also a parallel, if perhaps slightly later, trend in northern Europe towards a 2 × 8′ disposition in place of a unison and octave register.

It is in the first half of the 17th century that the strongest representation of harpsichords (and virginals) with extra keys for notes not normally tuned in mean-tone temperaments is found. This usually involved adding split sharps to provide a *d*♯ and an *a*♭, but other instruments were built with 19 notes to the octave, that is, with five alternative accidental keys and *e*♯ and *b*♯. This may have been a response to a fashionable interest in chromaticism, but it was also in order to overcome the limitations of the modulations and consonances possible within mean-tone tuning systems. Thus, some transpositions would have been facilitated by the addition of these extra notes.

For continuo purposes it is understandable that the harpsichord should have been equipped with an extra 8′ register (i.e. 2 × 8′ instead of 1 × 8′, 1 × 4′) in order to reinforce the fundamental. That this is a plausible interpretation is suggested by the fact that some 3 × 8′ harpsichords were also built; what these instruments might have lacked in subtlety was presumably compensated for by the increase in the strength of the 8′ sound. Here it must be remembered that simply increasing the number of registers does not automatically contribute more volume: the ultimate source of energy which sets the strings in motion is the player's fingers. Diruta (1593) counselled that the voicing of harpsichords should be light in order that the player have control over his fingers and not be reduced to thumping out dance tunes.

Whilst there may be a difficulty in interpreting the significance of a large compass in the 16th century, large compasses can be taken at face value towards the end of the 17th century. There are some specifically Italian traits to note: the *G′/B′* to *c′′′* was practically a standard range for French instruments in the 17th century, but not for Italian harpsichords. Where it occurs at all, it is usually because it replaced a 50-note *C/E* to *f′′′* compass. Surviving instruments also indicate that a chromatic bass octave to *C* was never common in Italy. Russell (1959) has reported that an inventory of Cardinal Ottoboni's instruments listed eight harpsichords with a chromatic bass octave, but it seems unlikely that this is correct. The term that is used to describe the bass octave, *ottava stesa* (spread, literally 'extended', octave) was used quite generally and referred to several types of bass octave.

Although in the 17th century harpsichords were built with different specifications, the construction methods remained much the same. It can be noted, however, that many of the later instruments were more simply decorated than 16th-century harpsichords: soundboards often did not have a rose, keyboards lacked fine embellishments and elaborate inlay work on cases is rarely found.

Whereas 16th-century harpsichords were normally made with thin case sides and light mouldings and kept in separate decorated cases, another style of construction made its appearance in the 17th century: a thick-cased, painted harpsichord, possibly with cypress veneer and mouldings around the inside edge above the soundboard so that it seemed as if an instrument were inside a separate, outer case ('false inner-outer'). The idea has arisen that in the 16th century harpsichords were taken out of their cases and placed on tables in order to be played, but that by the beginning of the 17th century this was no longer the practice. However, it is quite clear that some of the earliest harpsichords have original outer cases in which the instruments were intended to remain while they were played, since a small door was made in the cheek of the outer case to give access to register knobs. Indeed, decorated outer cases were pieces of furniture and were intended to be seen as the harpsichords were played. Thus, the idea that harpsichords were taken out of their cases to be played has probably been overstated.

Few makers in the 17th and 18th centuries seem to have gone to the trouble to provide means for changing the registers. An obvious conclusion to draw is that contrasts of registers or volume were not an essential part of Italian performing practice. A few harpsichords had a single row of jacks with two tongues (one plucking left, the other right) so that it was physically impossible to disengage one register.

Some two-manual harpsichords were made by Italians. The 1690 harpsichord attributed to Cristofori and a harpsichord of around 1650 both have a 2 x 8′, 1 x 4′ disposition. The keyboards have no coupler and the 4′ is on the upper manual. A few three-manual harpsichords are known but these are all alterations or fakes. Naples was important as a centre of keyboard composition in the 16th century and the early 17th, yet few Neapolitan harpsichords have survived. Those that remain do not appear to be substantially different from instruments made in other parts of Italy.

There are some inventory references to instruments with an *ottava bassa* (literally, 'low' or 'bass' octave), but whether this should be identified with a separate register (with its own bridge) at 16′ pitch is not clear. However, the length of a Zenti harpsichord described in the Medici inventory, about 3.25 metres, leaves little doubt that *ottava bassa* could at least designate an instrument with an extra octave, even if there was no separate 16′ stop.

Some harpsichords were constructed with three 8′ registers and the technical realization of this is of interest: brass and iron scalings were used, with the result that either two soundboard bridges were necessary (with one nut), or, as in one known instrument, a combined soundboard bridge carried all three strings (with two nuts).

4. 18TH CENTURY

(i) France

The history of harpsichord making from the last decade of the 17th century into the 18th is largely an account of the rather rapid replacement of the relatively thin-cased instruments of the kind described above by a national variant of the thick-cased, long-scaled Ruckers design. Although a Louis Denis double of 1658 survives with a thick case and a long scale, the type did not become common until the 1690s. A Nicolas Blanchet double dated 1693, with an original range of *G′/B′* to *c‴*, has a pine case 15 mm thick with the surprisingly long scale of 36.2 cm; however, the case sides overlap the bottom and the moulding on the inside of the top edge of the case is in the 17th-century style. The 8′ bridge is of the massive Flemish design (although the slope is on the front rather than the back), while the 4′ bridge is Italianate with a moulding. From only 14 years later a large double harpsichord, a 1707 Nicolas Dumont, exists in original condition with a range of *F′* to *e‴*, having all the characteristics of a mature 18th-century French instrument. The reason for this rather radical shift in building style seems to have been the preference of musicians for the tone of Ruckers instruments. At the turn of the century increasing numbers of Ruckers harpsichords were finding their way to France where they received French keyboards and actions. A normal Ruckers transposing double with the range of *C/E* to *c‴* on the upper manual could accommodate the range of *G′/B′* to *c‴* without altering the string spacing on the bridge or the scale, simply by realigning the keyboards. A second choir of 8′ strings would usually be added, and the number of registers would be reduced from four to three, giving the normal French disposition (see fig.14). Often an extra note was squeezed into the bass to provide a split E♭ and two were added in the treble to extend the range to *d‴*. A less common type of transposer had chromatic basses, and such an instrument could be altered as above to a range of *G′* to *c‴* (see fig.7). Almost as many Ruckers or Couchet harpsichords with French keyboards and actions, but with unaltered cases and soundboards, survive from the first quarter of the 18th century, as do original French instruments of the period, amply attesting to their popularity. This popularity continued throughout the century, but the demand for an increased range altered the purity of design of these early *ravalements*.

Three harpsichord makers of this early period who have left important surviving instruments are Dumont and Blanchet, already mentioned, and Pierre Donzelague. Nicolas Dumont was admitted to the Paris guild of instrument makers in 1675. In addition to the 1707 double by Dumont there exist two others dated 1697 (in the Paris Conservatoire collection) and 1704 (privately owned in Paris). The 1704 one was so thoroughly rebuilt in the late 18th century that its original plan is obscure. The earlier one, however, was beautifully rebuilt in 1789 by Pascal Taskin and it is possible to determine its original design: a thick-cased long-scaled instrument with a range of *G′*/*B′* to *c‴*, or at most *G′* to *c‴*; this shows that the period between the 17th-century small instrument and the large 18th-century model is shorter than has been supposed. Nicolas Blanchet, who was admitted to the guild in 1689, founded the longest and most important dynasty of Parisian harpsichord makers, including his son François Etienne (*c*1700–1761), grandson François Etienne (*c*1730–1766) and Taskin. Seven harpsichords, three spinets and at least seven rebuilds of Flemish instruments by the Blanchets are known to survive. Pierre Donzelague went to Lyons from Aix-en-Provence in 1688 and became well known there for his work. There are two large and beautifully made double harpsichords by him: one in a London private collection, dated 1711, the other dated 1716, in the Musée des Arts Décoratifs, Lyons. They are the earliest-known French harpsichords with the full compass of *F′* to *f‴*. The majority of 18th-century French harpsichord makers were of the Parisian school, but there was a distinct though similar school in Lyons. 18th-century harpsichords from other parts of France are rare, and most of them are either archaic or are the occasional work of an artisan of another craft such as organ building.

Except for their size, the construction of these early 18th-century harpsichords was very similar to that of Ruckers. The framing was a bit heavier, especially the upper level braces which were more numerous. A horizontal brace was glued to the back edges of the upper belly rail in two Blanchet harpsichords of 1730 and 1733. This brace or 'T' section enormously stiffened the belly rail, and braces were run from it to the treble of the bentside strengthening this critical point. In some instruments the upper braces are set on edge and notched to the liners rather than lying flat under them. The lower braces of the 1707 Dumont have uprights or knees attached to their ends which are notched to the liners. Case sides were sometimes of pine rather than poplar which was invariably used by Ruckers. It should be noted that bentsides were always of poplar, and later lime, pine being difficult to wet-bend. Early in the century, the bentsides assumed a characteristic 18th-century shape with the curve concentrated in the treble and the remainder towards the tail, straight. The 1693 Nicolas Blanchet has a very incurved bentside, whereas an undated Nicolas Blanchet of the first quarter of the 18th century (perhaps in the first decade, since its case sides overlap the bottom) has a bentside only slightly so. The bentside of the 1707 Dumont is straight for half its length and those of the 1730 and 1733 Blanchets are straight for almost two thirds of their lengths. This shape continued in use in the Blanchet–Taskin workshop and was also used by Henri Hemsch and others; completely curved bentsides occur only occasionally later in the century.

The soundboard barring, especially in the first half of the century, was not nearly so standardized as Ruckers barring. The 1707 Dumont and 1733 Blanchet lack cut-off bars, and the ribs perpendicular to the spine extend to the 4′ hitch-pin rail. The 1730 Blanchet has a normal cut-off bar but, like the 1707 Dumont, it has two ribs crossing the 8′ section of the soundboard around the mid-section and tenor, and a third approaches the bass of the 8′ bridge from the 4′ hitch-pin rail. Ribs that cross the 8′ bridge or approach it from either the 4′ hitch-pin rail or bentside liner continued in use until mid-century. An Antoine Vater 1738 double and a 1746 François Blanchet have curved cut-off bars following the 4′ bridge; the latter is wide and flat like a smaller 4′ hitch-pin rail. The 1738 Vater and a 1742 Louis Bellot have ribs parallel to the spine crossing both bridges. All 18th-century French ribs observed were cut out where they passed under bridges.

The standard range from early in the century was *F′* to *e′′′*, but *G′* to *e′′′* was not uncommon. This is strange as the music of the period almost never exceeds the *G′* to *d′′′* range of the larger Ruckers type. The *F′* was used in one piece each by Rameau and

François Couperin in their solo harpsichord works, but it was not in general use until the 1740s. Neither Couperin nor Rameau employed *e'''* in their solo works. Dagincour used it in 1733 (*Pièces de clavecin*), but it was not often found until the *F'* was commonly written. The *e'''* seems to have fulfilled a French sense of order: the keyboards were balanced with one natural after a group of sharps at each end. *G'* to *e'''* instruments, such as the 1742 Louis Bellot and a 1748 Johannes Goermans, continued to be made almost to mid-century. During the 1720s some of the short-octave Ruckers conversions had chromatic basses to *G'* (often without *G♯'*) crowded into the existing cases.

The keyboards of these instruments continued the design of the previous century with a few stylistic changes. Fruitwood arcades replaced the carved trefoils on the key fronts, and the sharps, instead of being solid ivory or bone, were composed of a thin bone slip glued to a black stained block. In better workshops, such as that of the Blanchets, the sharps were tapered both in height and width, and the key-levers, seldom leaded, were individually carved to balance. The span of an octave remained small, averaging about 15.9 cm. The wooden jack-slides and guides were covered with punched leather which became the bearing surface, and the accurately made jacks were slightly tapered in width and thickness, fitting the slide only when at rest. These actions were light, quiet and repeated very quickly. The disposition of an 8′ and 4′ register on the lower manual, coupler, and an 8′ on the upper, with the 4′ between the 8′s and the lower 8′ plucking the longer string, seems to have been absolutely standard until the third quarter of the century.

In the 1740s the Blanchets' connection with the court began, and shortly after the middle of the century their firm became 'facteur des clavessins du Roi'. During this time, besides their maintenance work for the court, they became increasingly occupied with the rebuilding of Flemish harpsichords into large five-octave French instruments. Two other families should be mentioned: Johannes (Jean) Goermans (1703–77) and his son Jacques Goermans (*c*1740–89); and the brothers Henri and Guillaume Hemsch. About eight of the Goermans' harpsichords and six by the Hemsch family survive; five of Henri Hemsch's date from the decade 1751–61, a remarkable survival rate. Mid-century harpsichords continued the style of the earlier ones. Framing became a bit heavier and more sophisticated, the sides

17. Harpsichord by Pascal Taskin, Paris, 1769; see also fig.24

were a little heavier and Blanchet began to taper their thickness from the keyboard end towards the tail. Lime and pine were preferred to poplar. Soundboard barring became less experimental and settled into the Ruckers pattern. An *f'''* was added to the treble but later than musicians asked for it. A 1754 Hemsch ends with *e'''*, and although a 1756 Hemsch has the *f'''* it may have been added later, as was often done to earlier instruments. 1758 Blanchet keyboards in a 1680 Couchet clearly show the addition of the *f'''* while the keyboards were being made.

In 1766 Pascal Taskin inherited the Blanchet workshop and from then on dominated Parisian harpsichord making; only a few single examples of other late makers survive. Taskin continued the Blanchet tradition using and refining their plan. The framing, ribbing and 4′ hitch-pin rail were beautifully rounded to save weight. He increased the taper of the sides and returned to the curved cut-off bar. The action of his instruments was even more refined than that of his predecessors.

After the mid-century several additions to the standard disposition began to appear. Although common on Ruckers instruments, the buff stop was rare in France in the first half of the 18th century. Neither the 1746 Blanchet nor the 1754 Hemsch originally had one. When it did come into use, it became almost universal, and double buff stops (which can buff either 8′ string) were not unusual.

The three-register disposition sufficed on French instruments until the 1760s, when a fourth register having jacks fitted with plectra of soft buff leather was added behind the other three as a special solo stop. This addition is credited to Taskin in various writings of the period, although it is also ascribed to the organist and composer Claude-Bénigne Balbastre (who is known to have had a harpsichord fitted with *peau de buffle* plectra in 1770) and to a certain 'M. de l'Aine', who in 1769 announced an instrument fitted with leather plectra.

In the late 1750s French harpsichords began to be equipped with a variety of foot- or knee-operated devices for producing crescendo effects and for changing registers without taking the hands from the keyboard. An instrument with a mechanism of this type was offered for sale by one Wittman (or Wetman) in 1758, and several other makers subsequently announced mechanisms of their own. The only examples to survive in any number are made by Taskin and Joachim Swanen, and employ a system (reportedly devised by Taskin in 1768) activated by five or six pommels that can be raised or lowered by the action of the knees. When raised, three of these act to disengage, respectively, the lower-manual 8′, the 4′ and the *peau de buffle* registers; a fourth pommel raises the *peau de buffle* jacks when they are not in use in order to keep the touch as light as possible; a fifth pommel, when gradually raised, disengages first the 4′, then the lower-manual 8′ and then the upper-manual 8′, permitting a gradual decrescendo from the *forte* of all registers sounding at once to the *pianissimo* of the *peau de buffle* alone; an additional sixth pommel, found on a small number of instruments, acts to slide the upper manual inwards, thereby coupling it to the lower manual so that on these harpsichords the only register to be controlled by hand is the buff stop. It should be emphasized that these devices came into being before the piano achieved any popularity in France and, accordingly, they do not represent an attempt to meet the challenge of the piano with its inherent ability to produce dynamic nuance; rather, they appear to have been an independent manifestation of the interest in dynamic expression that had led to the invention of the piano in the closing years of the 17th century.

A great portion of the energies of late French harpsichord makers appears to have gone into the massive rebuilding of older harpsichords, especially those of the Ruckers family. Since a rebuilt Ruckers harpsichord was worth several times as much as a new instrument in 18th-century Paris, such a diversion of the makers' efforts from building new instruments was clearly justified on a financial basis and it led not only to the most elaborate sort of rebuilding, including the conversion of narrow-range single-manual instruments to wide-range doubles and the building of new harpsichords around the soundboards of old virginals, but also to outright faking of new instruments to make them look like rebuilds. But as the rebuilding was intended to update earlier instruments to current musical requirements and not to preserve their antique qualities, the sound of a Ruckers harpsichord rebuilt by Taskin represents 18th-century Paris rather than 17th-century Antwerp.

Blanchet and Taskin were famous for their work in this vein, and they applied to it all the ingenuity and craftsmanship found in the instruments they built in their own names, producing neither crude enlargements in which extra notes are crammed into the bass (in effect sliding the keyboard towards the treble, thereby disastrously shortening the scaling) nor such dubious expedients as the jointing of extensions on to the wrest plank and belly rail. Rather, they used a wide variety of slightly differing techniques, determined by the nature of the original instrument. Of these, the most subtle and ingenious involved rebuilding the spine, in addition to the usual extending of the bentside and bridges and replacement of the cheekpiece, wrest plank and belly rail with new ones of appropriate length. The front of the original spine was cut off at the belly rail and the top was cut down to the level of the soundboard. A tapered layer of new wood of the same size would then be added on the outside of the cut-down original spine; then a wholly new spine of the same height as the rest of the case, and long enough to reach the front of the instrument, would be glued on to the outside of the tapered piece. The result was simultaneously to provide more room at the front of the instrument for additional bass keys and to rotate the entire body of the instrument anti-clockwise with respect to the strings. This rotation, in turn, had the effect of lengthening the scaling to compensate for the shortening produced by the addition of the new notes in the bass.

The plan of the average 18th-century French harpsichord more nearly follows that of the chromatic rather than the short-octave Ruckers transposer. This design had shorter tenor scaling to keep the tailpiece from becoming too wide in a wider instrument of the same length. The French tonal ideal around 1700 was that of Ruckers, but making larger instruments resulted in a grander and smoother tone. Although the 'presence' and immediacy of a small instrument were lost, the sound was no less transparent. As the century passed, the tone grew more complex and less direct; nevertheless even the late Taskins never lost the balance between attack and sustaining power that permits cleanness of articulation. The declamatory style of French keyboard music from the 17th century to the Rococo period required this sensitivity to articulation, and their harpsichords met the demand well.

Whether new or rebuilt, a French 18th-century harpsichord was often a major piece of decorative furniture. The soundboards were painted with flowers in a more sophisticated style than the Flemish, the cases were painted or lacquered in any of a variety of fashionable styles, and the instruments were equipped with elaborate six-, seven- or eight-legged bases often carved and gilded in one of the royal styles. Simpler instruments were painted in one or two colours, panelled with gold bands and mouldings and fitted with less elaborate bases but still in one of the royal styles (see fig.17). Despite the use of walnut and marquetry in 17th-century harpsichords, and the superb quality of veneered furniture in 18th-century France, French harpsichords seem never to have been veneered.

(ii) England

In England the standard 18th-century national type seems to have crystallized in the work of Hermann Tabel (*d* 1738), a builder, trained in Antwerp, who moved to London in about 1700. Both of the makers whose firms dominated English harpsichord building in the 18th century, Burkat Shudi (1702–73) and Jacob Kirckman (1710–92), worked in Tabel's shop and both built instruments strikingly like the sole surviving example of Tabel's work, a double-manual harpsichord dated 1721 (see fig.15).

A typical Shudi or Kirckman double has a 2 × 8′, 1 × 4′ disposition, with a dogleg jack and a row of lute-stop jacks controlled by the upper manual and a buff stop that on a Shudi acts on the lower-manual 8′ strings, and on a Kirckman on the dogleg 8′ strings (see fig.18, p.80). Their cases are made of oak, and are veneered in walnut with sycamore in early examples and mahogany with satinwood in later ones. The instrument is supported on a trestle stand with four legs, which vary throughout the 18th century from turned George II to square Chippendale; occasional special examples have rather ungraceful cabriole legs curving outwards from the level of the trestle's lower stretchers. The soundboards are not decorated with paintings, and Shudi soundboards do not have a gilded metal rose; the barring and case bracing are rather like those of a Ruckers harpsichord. Like the bottom, all the braces are pine. The lower ones are not as tall as in a Ruckers instrument; there are only two transverse bottom braces in addition to the lower belly rail, but these are supplemented by a diagonal brace running along the bottom from the intersection of the rear brace and the bentside to the centre of the forward brace. In addition, there are two longitudinal braces running upwards from the front bottom brace to the upper belly rail. The upper-level braces are more numerous than on a Ruckers harpsichord, where there are three set nearly parallel to one another and a slightly oblique angle to the spine. In a Kirckman or Shudi harpsichord there are four such braces which, however, are set vertically rather than flat, so that they bear on the face of the lining rather than merely being nailed to its underside. These four are supplemented by a fifth, heavier one, that passes from the bentside to the upper belly rail in the crucial treble area. The construction of a single-manual Kirckman or Shudi is identical with that of a double, except that while Shudi's singles often include a lute stop, Kirckman's singles seldom have one. Occasional examples of singles by both makers lack the 4′ as well.

Except in matters of decoration, these instruments changed little throughout the century, except for a shift in the plucking-points of Shudi harpsichords after 1770 that produces a rounder and less incisive tone in the later instruments (a change in line with the occasional substitution of leather for quill plectra in the

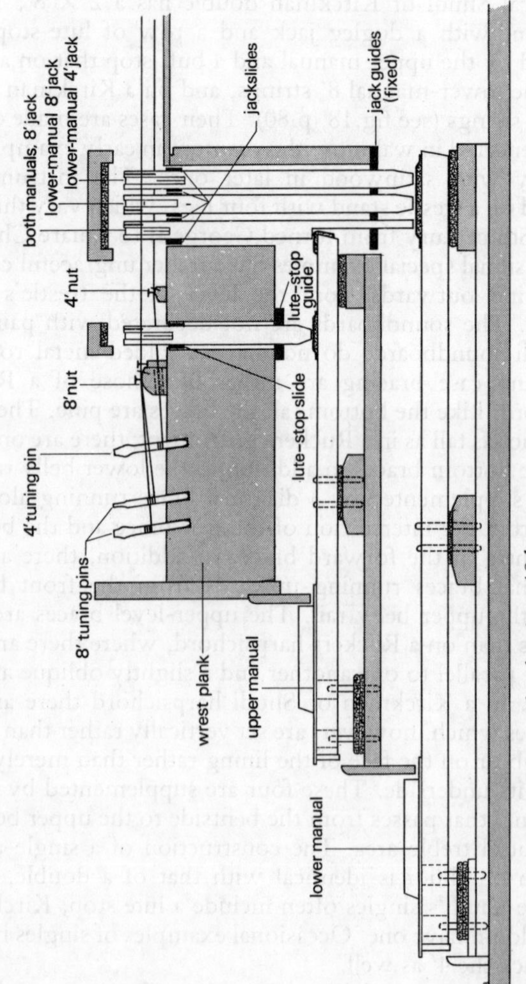

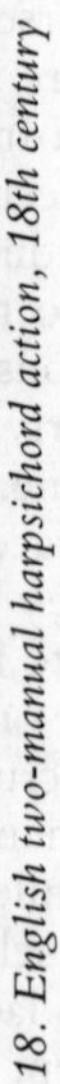
18. English two-manual harpsichord action, 18th century

lower-manual 8′ jacks), and the addition of the pedal-operated mechanisms described below.

Tabel's five-octave F' to f''' keyboard had lacked the $F\sharp'$ (presumably for reasons of visual symmetry), and Kirckman and Shudi, like other English builders, followed this practice until about 1780, when the $F\sharp'$ was included as a matter of course.

A minor difference between Shudi and Kirckman harpsichords concerns the arrangement of the stop-knobs in two-manual instruments. On Shudi double-manual harpsichords the three stop-knobs at the left side of the nameboard control (from left to right) the lute stop, the 4′ and the buff stop, whereas in a Kirckman the order is buff stop, lute stop and 4′; both have 8′ stops located at the right side of the nameboard with the dogleg controlled by the left-hand knob and the lower manual 8′ controlled by the right-hand knob. As a result of this arrangement, one can rapidly engage whichever of the unison stops may temporarily have been disengaged simply by squeezing the knobs together. Although Kirckman is known only once (1772) to have built an instrument with a range greater than five octaves (a double with a range of F' to c''''), Shudi regularly made instruments with a range of C' to f''', of which 12 dating from 1765 to 1782 have survived.

The tone of a Kirckman or Shudi harpsichord is enormously rich and powerful. As is true of many English and Hamburg harpsichords made in the second half of the 18th century – that is, after the great age of harpsichord composition – the sound of these instruments sometimes tends to call attention to itself rather than merely serving as a vehicle for projecting the music, a quality that may in abstract terms be viewed as a defect despite its splendour.

Beginning no later than the early 1760s, English harpsichords were customarily fitted with crescendo devices. The so-called machine stop disengages the 4′ register and then the front 8′ register as a pedal is depressed (on double-manual instruments, since the disengagement of the front 8′ register would silence the upper manual, it simultaneously engages the lute stop): thus, when the pedal is fully depressed the registration on the upper manual of dogleg 8′ is replaced by lute stop, and that on the lower of dogleg 8′, lower-manual 8′ and 4′ by lower-manual 8′ alone. In both single- and double-manual instruments, the machine stop can be disengaged when desired to permit normal hand-stop operation. By 1766, the machine stop was supple-

19. Harpsichord by Shudi and Broadwood, London, 1775, showing the swell in the open position

mented by a second crescendo device, the swell, which enabled the performer to open either a section of the harpsichord's lid (if not already raised) or a series of louvres covering the soundboard (see fig.19). The two devices used in conjunction with one another produce a surprisingly wide and effective crescendo, beginning with the *pianissimo* of the lower-manual 8′ alone with the lid or louvres closed, followed by the successive addition of the front 8′ and the 4′ and finally the gradual opening of lid or louvres to permit the *fortissimo* of the full harpsichord.

(iii) Italy

In the 18th century the plainest instruments of all Italian harpsichord making are found, with many false inner-outer or non-inner-outer types. There are indications that instrument building in other countries may have influenced some Italian makers. A clear example is the 1792 Sodi harpsichord: with its rounded tail and sloping cheeks it clearly resembles the Stein-type fortepiano. In the 1726 Cristofori harpsichord (Musikinstrumenten-Museum, Karl-Marx-Universität, Leipzig) no attempt was made to give the illusion of a harpsichord in a case, as there are no mouldings or cypress around the inside of the instrument. Only the characteristic Italian keyboard with boxwood naturals gives a clear indication that this is an Italian instrument. The nameboard is no longer the thin, removable Italian-type, but instead is permanently affixed in the manner of some 17th-century French harpsichords. Roses, where they occur, were relatively crudely made in comparison with the fine work found in 16th-century instruments.

By the beginning of the 18th century most of the alterations to 16th-century instruments (see §3(iv) above) had already been carried out. Goccini, who is known by some of his own instruments, altered in 1704 an Alessandro Trasuntino harpsichord (made in 1530; privately owned) as his signature on the new keyboard shows. Bartolomeo Cristofori is better known as the inventor of a piano escapement, but some of his work in repairing or altering instruments has been documented in the Medici archives. The invoice dated 10 February 1693 'For rebuilding a harpsichord of Domenico of Pesaro . . . including remaking the keyboard and bridges' may refer to the 1554 Dominicus Pisaurensis harpsichord (in the Paris Conservatoire). In any event, the new bridges fitted to this instrument correspond to Cristofori's own. This alteration yields some interesting clues about the pitch of instruments at this time: although the scale that would have resulted from Cristofori's keyboard modifications would have been 29 cm at *c''* (with the original Dominicus bridges), Cristofori chose to remove the bridges in order to provide a scale of 26 cm. The 29 cm scale was only a few millimetres longer than that which he used himself in several of his own instruments, so the new, 26 cm scale was a deliberate choice. Some 18th-century instruments have scales of about 26 cm, which would come to about *a'* = 440 Hz when strung with brass wire.

Recent research on harpsichords of all countries has attempted to elucidate the ways in which instruments were strung, in order better to understand the maker's intentions and provide more faithful restorations. This work has proceeded on the one hand by an examination of scalings and detailed stringing schedules found on instruments or documents (see O'Brien, 1981) and by archival work on the methods of wire manufacture (see Gug, 1984, 1986). Even without an exact knowledge of wire sizes, it is possible to see from the gauge numbers marked on some instruments that there was considerable agreement among some 18th-century makers about the wire sizes to be used. Hubbard (1965) supposed that the light construction of Italian inner-outer harpsichords required strings having less tension than thick-cased, false inner-outer instruments. A study of gauge numbers on some Cristofori instruments has shown that he did not string on the principles suggested by Hubbard (see Wraight, 1982).

Cristofori's work as a harpsichord maker also shows an inventive streak; indeed, it seems as if he needed new technical challenges in each instrument. There are simpler ways of making a 2 × 8′ virginal than the one Cristofori chose, but the graceful symmetry and delightful shape of his design wins admiration. One of his harpsichords has three registers at three pitches: 8′, 4′ and 2′. Although the instrument is mostly strung in brass, there is a break in the scaling in the treble with extra bridges set further apart for iron wire. Thus, the longer, iron scaling is used to make space for the three registers; if the instrument had been strung throughout in brass, then the treble ends of the bridges would have been on the registers, clearly an impossible arrangement.

Although wide compasses became common towards the end of the 17th century, the smaller instruments were not entirely abandoned. Some *C/E* to *c'''* compasses were still made in the

18th century; it is likely that these were intended for simple vocal accompaniment. Some of the latest instruments were made with compasses from *F′* to *g‴*. Domenico Scarlatti's popularity, which extended throughout Europe, may have influenced the Italians in building such a wide range. The last-known Italian harpsichord (in the Musikinstrumenten-Museum, Leipzig) was made in 1792 by Vincenzio Sodi. As in other countries, some Italian makers (such as Cresci) transferred their attention to the fortepiano when the harpsichord was no longer in demand at the end of the 18th century. Right to the end, Italian harpsichords were made with a 2 × 8′ disposition, and few two-manual instruments were produced. Furthermore, the 2 × 8′, 1 × 4′ disposition so often used in other parts of Europe was rarely used by Italian makers.

(iv) Germany and other European countries

Compared to the number of surviving 18th-century harpsichords from Italy, France, England and the Low Countries, there are progressively fewer from Germany, Scandinavia, Portugal and Spain, and hence progressively less information is available concerning the character and development of the instrument in these areas. This is specially regrettable since Germany and Spain in particular produced so much harpsichord music of interest. Moreover, Germany produced at least two distinct schools of harpsichord making in the 18th century, of which only one – the Hamburg school – is represented by an appreciable number of surviving examples (see fig.20, p.86).

The Hamburg school included a substantial number of builders, of which the members of the Hass and Fleischer families are the most notable. The earliest surviving Hamburg instrument, a single, built in 1710 by J. C. Fleischer, already has all the basic characteristics of this school, the products of which, among all 18th-century harpsichords, seem most clearly to derive from the intermediate type. Flemish influence is evident only in the use of softwood for the case and in the typical north European 18th-century harpsichord scaling of 34 cm. The instrument has the curved tail typical of Hamburg harpsichords, and its case bracing is accomplished by full-depth members crossing the case from the bentside to the spine with no upper-level bracing. Intermediate-type influence is also evident in the coupling system used on Hamburg two-manual instruments, in which the lower manual is moved while the upper one remains

20. Harpsichord by Christian Zell, Hamburg, 1728

fixed. In these instruments, however, coupling is accomplished in an unusual fashion. Short doglegs are provided for the upper-manual 8′ jacks and, when the lower manual is pushed inwards, small padded blocks on the lower-manual keys are positioned under the doglegs so that the upper-manual jacks are then lifted by the lower-manual keys without the upper-manual keys having to be moved as well.

The harpsichords of the Hass family are noted for their complex dispositions and are the only 18th-century instruments known to include with any regularity a 16′ register. Two of the three surviving instruments so provided also have a 2′ register in addition to the 16′, 4′ and two 8′ registers. (One, of about 1770, is at Yale University, and the other, dated 1740, is in the Puyana Collection, Paris; there is also a Hass harpsichord of 1723 in the Musikhistorisk Museum, Copenhagen, with three sets of 8′ strings plus a 4′ set.) No two of the three instruments with a 16′ register are exactly alike, but their 16′ strings are arranged in the same ingenious fashion. Inside the case a low curving rim is attached to the deep frame members and follows the line that the bentside of a normal instrument would take. This rim serves as a hitch-pin rail for the 8′ strings. Beyond it and at a slightly higher level there is a completely separate soundboard for the 16′ bridge, and the 16′ strings are then hitched to the pins driven into the lining of the bentside along the far edge of this separate soundboard. As a result, the 16′ bridge does not have to be pierced to permit the 8′ strings to be hitched at the bentside, and the layout of the 8′ and 4′ strings, which still comprise the basic core of the harpsichord, is undisturbed. The addition of this extra soundboard to an already full-size instrument means that such a harpsichord may exceed 275 cm in length.

The surviving instrument with a 16′ stop but without a 2′ stop, which was made by Hieronymus Albrecht Hass in 1734, is equipped with a close-plucking lute stop (arranged like that on an English harpsichord to pluck the same string as the regular upper-manual 8′) as well as buff stops on both the lower-manual 8′ and the 16′; but the next surviving instrument by this maker, built in 1740, the only unquestionably genuine three-manual harpsichord still in existence, manages to include a 2′ stop as well. The upper two manuals of this instrument provide the same resources as an 18th-century English double-manual harpsichord: a lute stop on the upper manual, a dogleg 8′ register played by both the upper and the middle manuals, and a

4′ and a second 8′ playable on the middle manual only. The doglegs in this instance reach down to the middle manual, and there is no coupler between these two keyboards. The 16′ and the 2′ are confined to the lowest manual, which (like the keyboards of some chamber organs) can be pushed entirely into the case like a drawer for playing on only the 8′ and 4′ registers, but can be pulled forward part of the way so as to play the 16′ and 2′ by themselves, or further forward to permit all the registers except the lute stop to sound at once from the lowest manual.

This arrangement is certainly the most useful possible way in which to include the 16′ and 2′ stops if they are to be included at all. The problem with the 16′ stop, even on an instrument like this (in which the 16′ strings have their own bridge permitting them to be of adequate length and preventing their loading down the 8′ bridge), is that although the 16′ lends enormous solidity and gravity to the bass it tends to produce an undesirable thickening of the texure and a possible effect of downward transposition by an octave when engaged to double all the musical lines at the lower octave. The 2′ seems to have been added primarily to brighten an ensemble already made too dark or muddy by the inclusion of a 16′ stop; so the two belong together and may properly be isolated on a manual of their own. When this is done, the left hand can play on this lowest keyboard while the right hand plays on the middle manual. With the two manuals coupled, the 16′ and 2′ can thus strengthen the bass without thickening the texture by unnecessary doubling of the parts played by the right hand. Moreover, by pulling the lower manual only partly out so that it will not be coupled to engage the middle manual, the 16′ stop becomes available as a solo register either to be used in contrast to the tutti of two 8′ registers and a 4′ on the middle manual, or in dialogue with the lute stop on the upper manual or the back 8′ on the middle manual.

The remaining Hass instrument equipped with a 16′ stop has only two keyboards, but compensates for this by having two rows of 2′ jacks (both playing the same strings), one on the upper manual and one on the lower. This instrument bears the signature of H. A. Hass's son Johann Adolph, and a date of 1710, at which time the older Hass was only 21, so that his son could not possibly have made the instrument. Additionally, the fact that the instrument has a fully chromatic *F′* to *f‴* keyboard (rather than the *G'* to *d'''* of 1734, or the *F'* to *f'''* – lacking the low *F*♯' – of the 1740 triple) suggests that it was probably made closer to 1750 than to 1710. As on the 1734 double, buff stops are provided for the lower-manual 8' and the 16', and the lower-manual 2' – like that on the 1740 triple – extends only from *F'* to *c''*. This curtailment is necessary because, even with the narrow jackslides used on these instruments, the gap between the wrest plank and the belly rail required for five slides (and thus the minimum distance between the 2' nut and 2' bridge) must be so wide that no string stretched across it could be tuned appreciably higher than the *c''''* equivalent to *c''* at 2' pitch. This also explains why the sixth slide carrying the jacks of the upper-manual 2' on the (incorrectly dated) 1710 instrument has to be ended even earlier and, in fact, goes only to *b*.

The ingenuity of these instruments is matched by the care and craftsmanship with which they were made: even the tops of the jacks, for example, are finished with a small 45° bevel on each of the upper corners. The keys are often covered with tortoiseshell, and the painted or lacquered décor is correspondingly elaborate, chinoiserie being the most common style.

In German harpsichords the increase in compass proceeded in a slightly different manner from that of the French. The *G'*/*B'* short octave occurs only rarely in German harpsichords (e.g. a single-manual instrument by Christian Vater of Hanover, dated 1738); other examples descending below *C* are either fully chromatic from *G'* or *F'*, possibly omitting *G*♯' or *F*♯', as in the harpsichords by Michael Mietke at Schloss Charlottenburg. The range was expanded downwards to *F'* and upwards to *d'''* almost simultaneously, though at least three surviving instruments with a *C* to *d'''* compass date from the two succeeding decades. The five-octave *F'* to *f'''* compass typical of harpsichords of the second half of the 18th century is found as early as 1722 in a Saxon instrument (by Johann Heinrich Gräbner of Dresden, now at the Villa Bertramka, Prague; a transposing instrument, its keyboard can be shifted down a semitone). But the earliest surviving Hamburg example with such a range is dated 1740 (Puyana Collection, Paris). It is interesting to compare this progression with that found in datable German harpsichord music from before 1750. The 12 suites published by Johann Mattheson in 1714 keep within a *C* to *c'''* compass. Much of J. S. Bach's music also fits within such a range, but a number of his later works ascend to *d'''*, as, for instance, the partitas in C minor (BWV826, published in 1727) and G major (BWV829, published in 1730). Handel's first collection of suites, issued in 1720, requires a *G′* to *c‴* compass, while the second collection of 1733 ascends to *d‴*. Gottlieb Muffat's *Componimenti musicali* (Augsburg, *c*1739) demands the low *F′* but does not rise above *c‴*. Bach's French Overture (BWV831), published in its definitive B minor form in the second part of the *Clavier-Übung* (1734), regularly descends to *G′* but avoids the *F♯′* which would result from the literal transposition of the earlier C minor version of this dance suite. The fourth part of the *Clavier-Übung* (*c*1742), the famous Goldberg Variations (BWV988), also ranges over a *G′* to *d‴* compass. But C. P. E. Bach in his two sets of sonatas *per il cembalo*, the 'Prussian' (WQ48, published in 1742) and the 'Württemberg' (WQ49, published in 1744), wrote for a *G′* to *e‴* compass.

The sound of the Hamburg harpsichords does not usually fulfil the expectations raised by the ingenuity and craftsmanship manifested in their design and construction. Their tone is exceedingly bright and penetrating, and the upper harmonics seem to sustain rather longer than on other instruments of the 18th century. As a result, the sound lacks some of the transparency of other 18th-century instruments, a characteristic that is intensified in the examples with a 16′ stop, where the sub-octave doubling tends to make the musical texture even more opaque. Similarly, the addition of the 2′ stop, while brightening the tone and increasing pitch definition in the extreme bass, only contributes to the total thickness when it is used in performance. Both of these stops (and to some extent the entire concept of harpsichord sound apparent in these instruments) reflect an organ-based rather than a harpsichord-based orientation (for an opposing view see Williams, 1971).

The decoration of two surviving harpsichords attributed to Michael Mietke of Charlottenburg suggests a date no later than 1713. The one-manual instrument has two 8′ stops. Its original compass of *G′*,*A′* to *c‴* was later extended to *F′*,*G′* to *e‴*. The two-manual harpsichord has one 4′ and two 8′ stops with a coupler. Its range of *F′*,G′,*A′* to *c‴* was later altered to *F′*,*G′* to *e‴*. Neither instrument has a buff stop. It is known that J. S. Bach was sent from Cöthen to Berlin in 1719 to take delivery of a Mietke two-manual harpsichord for the Duke of Anhalt-Cöthen. Both instruments, now at Schloss Charlottenburg, have keyboards with a narrow octave span, even slightly smaller than on most French harpsichords, and keyheads almost as short as the French type. Bach is known to have preferred keyboards of such dimensions.

No undoubtedly authentic instrument by Gottfried Silbermann, Bach's friend, survives, but there are a number of spinets by his nephew Johann Heinrich Silbermann. Other extant instruments from outside Hamburg include a small group from Dresden, built by the members of the Gräbner family and their circle; two highly unusual harpsichords combined with pianos, by J. A. Stein; a Silesian example; and a few anonymous ones. These instruments are all significantly different from those of the Hamburg school, and they show their affinities with the intermediate type in rather different ways. The soundboard barring includes transverse ribs running under the bridges, and the cases are made of hardwood or are veneered rather than being painted, but round tails are rare. With the exception of the harpsichord included in one of the Stein combination instruments, none of these includes a 16′ stop, and this register must be considered as something of a Hamburg speciality. The tone of the Gräbner instruments is less brilliant and harsh than that of the Hamburg harpsichords, and has instead a characteristic dark reediness that is unlike the sound of the harpsichords from other countries.

Austrian 18th-century harpsichords are very rare. Boalch (1956) cites just one, signed 'Johann Leydecker, Vienna 1755' (Schloss Eggenberg, Graz). The Mozart household is known to have contained a large two-manual harpsichord by Friederici of Gera in Saxony, rather than one by a native Austrian or Bavarian builder. Evidence as to the extent of harpsichord making in Austria and Bavaria, as in Spain, is too sparse to justify drawing any definite conclusions. The same holds true for the older countries of the Empire, except for those areas of Italy ruled from Vienna. Virtually nothing is known of any 18th-century harpsichord building in what is now Czechoslovakia, Poland and Hungary, although there are indications that the craft had been practised there during the 16th and 17th centuries.

The only catalogued Danish 18th-century instruments of the harpsichord family that survive are a small virginal of 1762 (Rosenborg Castle) by Christian Ferdinand Speer, a Silesian émigré in Copenhagen, and a one-manual harpsichord of 1770 (Musikhistorisk Museum, Copenhagen) by Moritz Georg Moshack, a Copenhagen maker. It is known that there was an active trade in importing harpsichords from Hamburg during the century, and indigenous instruments were probably of similar design.

In Sweden, a number of instruments and some secondary evidence indicate that harpsichord making flourished during the 18th century. A two-manual five-octave harpsichord in the Stockholm Nordiska Museet, dated 1737, is signed by Philipp Jacob Specken, who had learnt his craft in Dresden before moving to Stockholm. Niels (or Nicolas) Brelin, a clergyman, is known to have built an upright harpsichord (clavicytherium) in 1741 with eight registration pedals. A contemporary sketch printed in the proceedings of the Swedish Royal Academy shows that it had a five-octave compass and that its disposition included a 4′ stop. Brelin is said to have made two trips abroad to study instrument building, but where and with whom he worked is not known. The harpsichord signed 'Johannes Broman, Stockholm 1756' (Musikmuseet, Stockholm) is a five-octave two-manual instrument, 3.6 metres long and similar in construction to a Hamburg harpsichord (including a double-curve bentside). It has a 4′ stop and three unison choirs (and the arrangement of the strings makes it clear that the third 8′ was never intended as a 16′ stop). The two-manual five-octave instrument signed 'Gottlieb Rosenau, Stockholm 1786' (Musik-historisk Museum, Copenhagen), while also similar in style to contemporaneous Hamburg instruments, is of relatively normal length, 276 cm.

Few 18th-century harpsichords from the north Netherlands (corresponding approximately to the modern Kingdom of the Netherlands) are recorded as extant: two instruments made in Amsterdam in the 1760s, a 1787 instrument from Leyden, and from Roermond a curious survival of 17th-century style, dated 1734. The Roermond instrument (in the Plantin–Moretus Huis, Antwerp) is an unusual two-manual harpsichord with a virginal filling out the space between the bentside and the extended cheekpiece. The harpsichord portion is reminiscent of an earlier transposing double after alignment. The compass is certainly the normal late 17th-century Flemish range, *G′/B′* to *c′′′*; the lower manual plays sets of 8′ and 4′ jacks, and the upper controls a dogleg 8′ and a second set of 4′ jacks playing on the same strings as do those of the lower-manual 4′. The virginal, with keyboard to the left, is of *C* to *c′′′* compass and is also archaic in lacking the low *C♯*. Johannes Josephus Coenen, as the maker signed himself, was a priest as well as the organist of Roermond

Cathedral, and seems to have made instruments in his spare time.

In sharp contrast a modern, large two-manual instrument boasting a 16′ stop was advertised for sale in Amsterdam just a year later (1735) by Rutgert Pleunis. This maker's extraordinary career as one of the most inventive keyboard instrument builders of his time was centred from 1741 in London (see fig.53 and Chapter Four, §10), where he was known as Roger Plenius. Unfortunately no instrument of either his Dutch or English period survives.

A harpsichord now at Leipzig, unsigned but with the intitials 'L.V.' in the rose, bears the date 1766 on its top key (*f‴*) and its place of origin, Amsterdam, on its lowest key (*G′/B′*). This single instrument is in the Flemish tradition, with the ends of the registers that protrude from the cheekpiece serving as stop-knobs for the two 8′ and the 4′ registers. A one-manual harpsichord of 1768, by C. F. Laeske of Amsterdam (private collection, New York), has two 8′ registers and one 4′, with a *C* to *f‴* compass. A harpsichord of similar disposition and compass, signed 'Dirk van der Lugt, Amsterdam 1700', disappeared from the Staatliches Institut für Musikforschung, Sammlung alter Musikinstrumente (the Berlin Collection), before World War II.

A harpsichord by Abraham Leenhouwer, a musician of Leyden (Gemeentemusuem, The Hague), a quite standard two-manual instrument of five-octave compass, disposed 2 × 8′, 1 × 4′, is remarkable not only for its very late date, 1787, but also for the archaic stop-knobs – extended register ends protruding through the cheekpiece. This feature, also found in the Coenen, 'L.V.' and Laeske harpsichords, seems to have survived longer in the north Netherlands than anywhere else.

In the south Netherlands (roughly equivalent to modern Belgium) a considerable number of instruments remains to substantiate the written record. In the early 18th century new harpsichords began to be made in the form that was characteristic of earlier instruments of the Ruckers type after they were enlarged in the late 17th century. Two 8′ stops rather than a single unison register were the rule. Two-manual instruments had either three sets of jacks (one each for the two 8′ and one 4′ choirs) or four, as in the former transposing harpsichords. In the latter case, the fourth set would be used either as a second 4′ stop playing on the upper manual (as in the Coenen harpsichord of

1734), or for a cut-through lute stop, plucking one of the unison choirs close to the nut. But quite a few simpler instruments continued to be produced, even in the late 18th century: Albert Delin of Tournai, for instance, seems to have done without a second manual or 4′ stop, although he was a builder of great skill and refinement, judging from his surviving ten or so instruments, dated 1750–70. In addition to making conventional harpsichords and spinets, Delin also produced clavicytheria that are outstanding both for their mechanical excellence and their rich sound, which – projected directly at the player – is quite overwhelming at first. Three examples survive (Berlin Collection, Brussels Conservatory and Gemeentemuseum, The Hague).

Jérôme Mahieu of Brussels was probably active before 1732, the earliest date recorded for him, for he died in 1737. He built harpsichords of both one and two manuals, generally with three registers (2 × 8′, 1 × 4′), but occasionally with only two, when he preferred the older 1 × 8′, 1 × 4′ disposition to the more modern 2 × 8′. The compass was either of 58 (*G′* to *e‴*) or 61 (*F′* to *f‴*) notes. (The 1732 Mahieu instrument with an apparent compass of *D′* to *d‴* reported in Paris in 1952 was presumably altered by a 19th-century restoration from the original *F′* to *f‴* range.)

Also active during the mid-18th century was Jacobus Van den Elsche of Antwerp. One instrument from his workshop has survived (in the Vleeshuis Museum, Antwerp). It bears the date 1763 and, save for its exceptionally sturdy construction, is a standard two-manual five-octave instrument of classic 2 × 8′, 1 × 4′ disposition. Another instrument (formerly in Berlin; destroyed 1945) was ostensibly dated 1710, seven years before Van den Elsche's entry into the Guild of St Luke, and signed to indicate that it was rebuilt in 1790 by Johann Heinemann, a reputedly blind harpsichord and lute maker of Antwerp; how with such a handicap he could have rebuilt the Van den Elsche instrument or made any of his own has not been explained. A one-manual harpsichord by Heinemann (Brussels) with a *C*/*E* to *d‴* compass, disposed 2 × 8′, is dated 1793; this would make it apparently the latest extant Flemish harpsichord, but the short-octave keyboard is strangely archaic, particularly in view of the date.

Members of the Dulcken family were distinguished harpsichord builders in the region during the 18th century. At least eight harpsichords by Joannes Daniel Dulcken are known. He was at first in Maastricht; in 1738 he settled in Antwerp, where he produced most of his harpsichords, and died in 1757 (instruments made in Brussels, and bearing later dates, are the work of his sons). Of a somewhat experimental turn of mind, J. D. Dulcken evolved an unusually long model of northern European harpsichord, his two-manual instruments being some 260 cm long. Occasionally he also made use of a singular type of construction with both an inner and an outer bentside. All his mature instruments are of five-octave compass, disposed 2 × 8′, 1 × 4′, often with a cut-through lute stop on the upper manual. Dulcken preferred to use a dogleg jack for the normal upper 8′ rather than a coupler (see fig.21, p.96). But since the lute register and the lower 8′ usually pluck the same choir, with the second unison strings sounding only when the dogleg 8′ is engaged, no dialogue of lower 8′ and lute stop is normally possible and the upper manual is limited to providing a softer sound contrasting with the tutti of the lower manual.

Joannes Petrus Bull, another German who settled in Antwerp, was apprenticed to J. D. Dulcken there. Four of his instruments have survived, dated from 1776 to 1789, all of five-octave compass and disposed 2 × 8′, 1 × 4′. Three are two-manual instruments. One of these, dated 1778, has most ingeniously wrought, very wide upper-manual dogleg jacks, with two tongues facing in opposite directions. These jacks can pluck either 8′ choir and thus a combination of 2 × 8′ is available on each manual, since the dogleg and the lute stop can be combined on the upper keyboard. But the lower 8′ jacks are fitted with *peau de buffle* plectra so that only the dogleg 8′ is available to give a normal quilled 8′ sound on the lower manual. Thus, as with Dulcken, no dialogue of a quilled lower 8′ and a lute stop is possible in the manner of the English double harpsichord. A later two-manual instrument by Bull (1789) lacks the double tongues in the dogleg upper-manual jacks; but it is so arranged that damper interference between the lower 8′ jacks and the dogleg upper 8′ even prevents using the upper keyboard as an echo manual.

Although in Switzerland some sparse records survive of harpsichord making as far back as the late 15th century, the only surviving instruments identifiable as Swiss date from the 18th century and come from the German-speaking area. There is no firm evidence that the craft ever took root in the other regions.

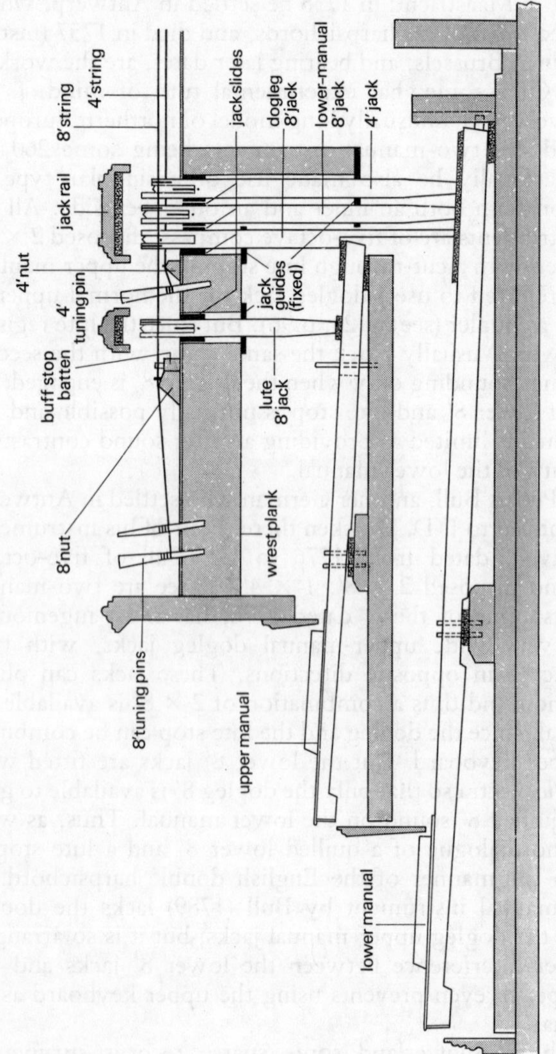

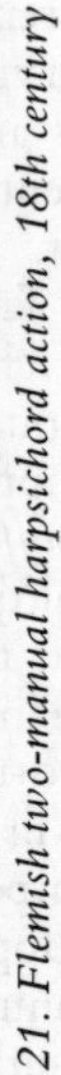
21. Flemish two-manual harpsichord action, 18th century

(A spinettino in the Schweizerisches Landesmuseum, Zurich, known to have been decorated in Stupan, Engadin, in 1722, is of uncertain origin and probably 17th-century.) Swiss harpsichords of the 18th century were probably similar in construction to the models produced in Strasbourg, particularly to those made in the Silbermann workshop. Peter Friedrich Brosi, a native of Swabia, was apprenticed to Silbermann before moving to Basle where he set up as an organ and harpsichord builder. A spinet signed by him (Schweizerisches Landesmuseum) is somewhat archaic for its date (1755), with a compass of *C* to *e'''*, a distinctly 17th-century type of dark walnut case and a black-stained stand of four heavy turned legs connected by a stretcher. A spinet of 1775 signed by his son, Johann Jacob Brosi, is closer in dimensions, compass (*F'* to *f'''*) and appearance to the late German type of instrument. An instrument by the Zurich craftsman Hans Conrad Schmuz, dated 1761, is in the Alstetten Museum. It is a single harpsichord of five-octave compass with two 8′ registers; the rather plain walnut case and simple turned legs strongly suggest provincial origins. An *ottavino* by his elder brother, Leonhard Schmutz [*sic*], was sold in Paris in 1924 on the dispersal of the Savoye Collection.

5. SINCE 1800

(i) 19th century

The Kirckman firm is said to have made its last harpsichord in 1809; the latest extant example is dated 1800. 19th-century restorers such as Tomasini, Danti and Fleury in Paris produced a few new instruments, and the harpsichord still appeared sporadically as a continuo instrument in oratorio and opera, and even as a vehicle for virtuoso pianists like Moscheles (1837) and Pauer (1861–7) to play in 'historical recitals'. But generally the traditions of harpsichord playing and construction slumbered in the 19th century; scholars, performers and public alike assumed that if Bach and Handel had known the modern piano in its iron-framed, cross-strung, double-escapement perfection, they would surely have preferred it to the 'deficient' harpsichords of their time.

In the mid-1860s the French virtuoso pianist Louis Diémer began to include in his recitals selections performed on the harpsichord, generally using a 1769 instrument by Pascal Taskin which was owned by the maker's descendants (but now at the Russell Collection in Edinburgh; see fig. 17 above). In 1882 this harpsichord was restored and subsequently borrowed for study by the Erard firm of piano and harp makers in Paris, with a view to resuming production of such instruments. Shortly thereafter the rival firm Pleyel also examined the Taskin harpsichord, and at the Paris Exposition of 1889 both firms displayed elaborately decorated harpsichords. Tomasini's more traditional harpsichord, based on an 18th-century instrument by Hemsch of Paris, was also shown (all three instruments are now in the Berlin Collection). Diémer presented with considerable success a series of historical recitals on the harpsichord during the Exposition.

The early revival Erard and Pleyel harpsichords – two-manual, five-octave instruments, disposed 2 × 8′, 1 × 4′ – are actually constructed more along the lines of English instruments of the mid- and late 18th century, such as Kirckman's and Shudi's, than those of Taskin instruments. Their framing, open at the bottom like that of the modern grand piano, is much heavier than that of 18th-century harpsichords, Erard's rather more so than Pleyel's. (Pleyel had been influenced by the piano to a greater extent in other respects such as scaling, soundboard ribbing and buttoned-on bridges.) While no metal bracing was used, the strings and bridges were far heavier than in antique instruments. The jacks were wooden, with traditional dampers. Erard used quill plectra in the lower 8′, but the other registers were leathered, as were all of Pleyel's (including the extra English-type cut-through lute stop which Pleyel added to the Taskin disposition). After initially opting for a combination of knee-levers and pedals, Erard changed to an instrument solely with pedals, such as Pleyel had made from the start. The elegant keyboards with ivory naturals and bevelled, canted ebony sharps are proportioned like those of the maker's pianos.

In London during the late 1880s Arnold Dolmetsch, a young French-born violin teacher who had trained at the Brussels Conservatory (where he had attended historical concerts with early instruments) and the Royal College of Music, London, began to present concerts of Renaissance and Baroque music. By 1890 he had acquired and made serviceable a Kirckman double harpsichord, an Italian virginal and a large German clavichord as well as a spinet and a small square piano. His concerts attracted a growing and influential circle of artists, writers and critics. In 1894 he constructed his first clavichord, and in 1896 at the suggestion of William Morris his first harpsichord, for display at the Arts and Crafts Exhibition in London. This was a one-manual instrument of *G′* to *f‴* compass, disposed 2 × 8′ with buff stop, and it so impressed the conductor Hans Richter that he engaged Dolmetsch, with the instrument, to accompany the recitatives in the 1897 Covent Garden production of Mozart's *Don Giovanni*. It was also used in Purcell performances in Birmingham. While antique instruments had served on rare occasions during the 19th century for continuo playing, this was apparently the first such use of a modern harpsichord. The revival in Britain owed much also to the efforts of A. J. Hipkins, a concert pianist, associate of the Broadwood firm and historian of keyboard instruments. In the 1880s and 1890s Hipkins gave lecture-demonstrations on 18th-century English harpsichords, using both his personal Kirckman and Shudi-Broadwoods from his firm's collection, and later the new Pleyel and Erard revival instruments.

In Germany and central Europe the harpsichord revival took hold more slowly. Almost from the first, moreover, a baleful influence made itself felt – the acceptance, as a model specimen, of a much-altered instrument that was falsely associated with Bach (harpsichord no.316 in the Berlin Collection) which had in fact been rebuilt earlier in the 19th century to replace a third 8′ register with a 16′ stop and to remove the 4′ register to the upper manual from its normal place on the lower keyboard. As early as 1899 a modern instrument based on no.316 was built by Wilhelm Hirl of Berlin for the Dutch collector D. F. Scheurleer, of The Hague. Other early German revival makers, such as Karl A. Pfeiffer (Stuttgart), Johannes Rehbock (Duisburg) and Georg Steingräber (Bayreuth), soon followed with their own versions of the spurious 'Bach' harpsichord. Even more elaborate and curious instruments were occasionally attempted in central Europe at this time, for instance a three-manual one by Seyffarth of Leipzig (1909; now in the Musikinstrumenten-Museum, Karl-Marx-Universität, Leipzig). The director of the Berlin Collection, Oskar Fleischer, published an article in 1899 summing up the aesthetics of the early harpsichord revival. He reported that the new Erard harpsichord had been seen and heard at the Vienna Music and Theatre Exposition of 1892 along with historical instruments from such collections as those of

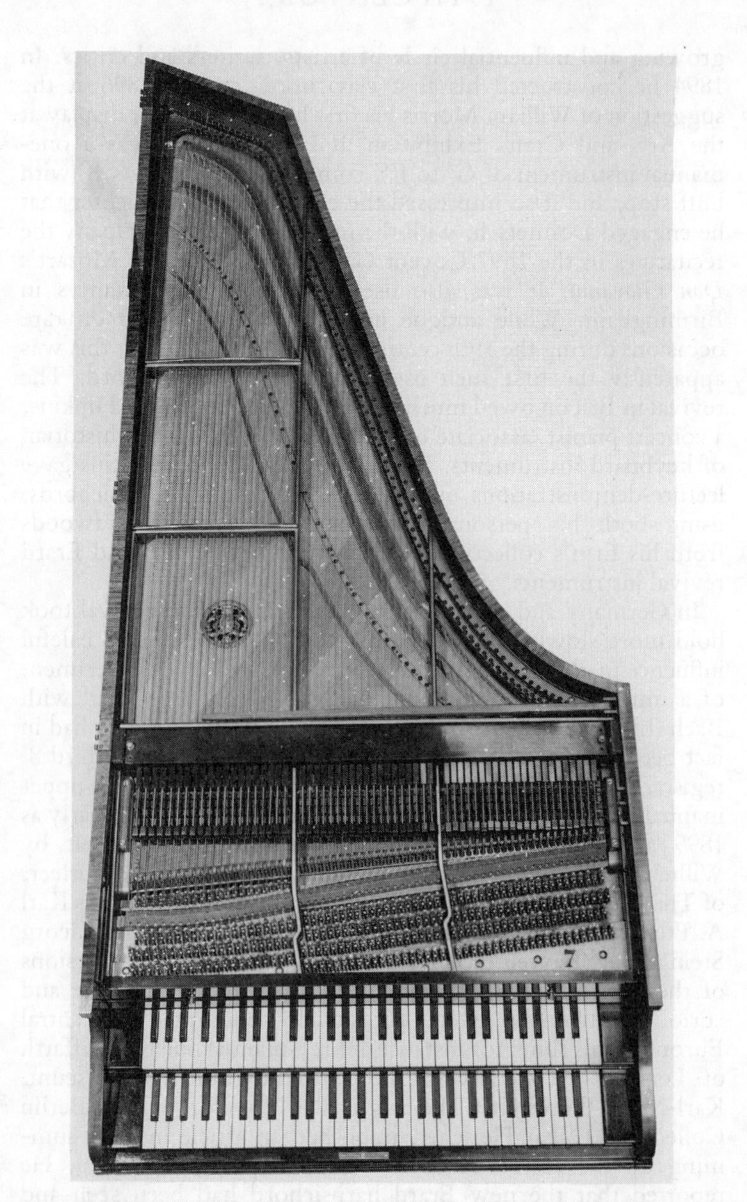

22. Harpsichord by Pleyel, Paris ('Landowska' model, first made in 1912)

Moritz Steinert (New Haven). He found the sound of the Erard 'hard, brittle and unsatisfying, quite apart from the lack of tonal combinations', and went on to praise Hirl's copy, allegedly faithful (save for a few small improvements) to the 'Bach' harpsichord which Fleischer had had acquired by the collection. He stressed, as a principle, variety of timbres and ease of changing registrations. The perfected modern harpsichord, with its pedals with half-hitches or special hand stops for dynamic variation, variety of plectra material and historically rare registers (16′ and cut-through lute), in addition to the basic 2 × 8′, 1 × 4′ disposition of the classic instrument, embodies the fulfilment of this ideal. In extolling such features, Fleischer particularly emphasized 18th-century music and the works of Bach and his contemporaries. (Apart from the Fitzwilliam Virginal Book, published in 1894–9, earlier keyboard music was largely unedited and little known.) The practicality and desirability of the 'Bach' disposition (lower manual: 16′, 8′; upper manual: 8′, 4′; plus buff stop and coupler) were assumed without question. Fleischer also raised some of the practical questions that continue to plague those concerned with presenting early music in the concert hall: whether the harpsichord can or should be capable of the level of loudness required to fill large auditoriums and balance modern string and wind instruments; and the best specification for an all-purpose harpsichord. At the time, when the shift from piano back to harpsichord was getting under way, there was as yet no concept of specialized instruments being specially suited to performing particular music.

(ii) 1900 to 1940

The Erard firm built harpsichords for only a limited period, but Pleyel continued their production. In 1912, at the urging of Wanda Landowska, the first modern harpsichord virtuoso of international renown, a new Pleyel model was introduced at the Breslau Bach Festival, and it was on this type of instrument that she performed, recorded and taught until her death in 1959. After 1923 the new Pleyel (see fig.22) also had an iron frame holding thick strings at high tension. The barring was almost identical to that of the modern grand piano, and the finely veneered case correspondingly heavily constructed. The touch depth and the dimensions of the five-octave keyboard were those of the modern piano. The cheekpiece and the spine were cut away in a delicately curved line to reveal the harpsichordist's

23. Harpsichord by J. C. Neupert, Bamberg ('Bach' model, first made in 1931)

hands playing on the keyboards. An extra set of overhead dampers was provided for the 16′ strings, and a highly sophisticated fine-tuning system was also fitted. The registers were controlled by seven pedals (but without half-hitches), disposed as follows – lower manual: 16′, 8′, 4′; upper manual: 8′, lute (*Nasat*) and buff; and coupler. The pedal action was largely negative (i.e. a pedal was raised to engage the register), a system that may have derived from the English 18th-century machine-stop pedal.

The arrival of this new Pleyel, first demonstrated in Germany, had a marked effect on harpsichord making in that country. Some makers now favoured the Pleyel disposition over that of the 'Bach' model, and the iron frame and generally heavier construction were taken up by such firms as Maendler-Schramm, a Munich workshop set up in 1906, and Neupert of Bamberg, a piano manufacturer (established 1868) that began harpsichord making at about the same time. The preference for the Pleyel disposition owed much to Landowska's influence as professor of harpsichord at the Berlin Hochschule für Musik, where she taught from 1913 to 1919, training an entire generation of harpsichordists. (In Germany, as elsewhere, very few harpsichords were made during World War I.)

From about 1930 most German harpsichord makers reverted to the 'Bach' disposition, and abandoned metal framing. Organist-harpsichordists, especially, had complained that an upper manual with only a single 8′ stop could not balance the mass of registers on the lower manual of the Pleyel-type instrument. Compromises of considerable mechanical ingenuity were offered by German makers, and later by some English builders as well: a 4′ stop normally played on the lower manual which could be coupled up to the second keyboard; 4′ strings playable by two sets of jacks, one for each manual; and two sets of 4′ strings, one for each keyboard.

From 1902, while the German revival was beginning, Arnold Dolmetsch had toured the USA extensively, presenting concerts of early music. In 1905 he was invited by the Chickering firm of piano makers in Boston to establish a department for the production of harpsichords, clavichords, lutes and viols. He accepted, and headed this department until 1910, when the firm's financial difficulties forced them to discontinue it. About 75 instruments of all types, including 13 harpsichords, were produced. These were two-manual instruments freely derived from a French 18th-century harpsichord used in Dolmetsch's concerts (the so-called 'Couchet-Taskin', actually by Goermans, dated 1764 and rebuilt in 1783–4 by Taskin; now in Edinburgh). The keyboards were back-pinned in the Taskin manner. While heavily cased, the Dolmetsch-Chickering harpsichords were lighter in construction than most other contemporaneous examples. The scaling was authentically 18th-century French but the ribbing of the soundboard, while light, was distinctly modern, crossing under the bridge. The tone, somewhat lacking in brightness, was nonetheless closer to the sound of antique instruments than any modern harpsichord had been. The lower 8′ was leathered and was provided with two sets of jacks, rather than half-hitches, to offer two dynamic levels. The upper 8′ was quilled, which rendered a combination with the lower 8′ less homogeneous, as in the case of the Erard model of 1889. The 4′ and the 16′, the latter added in 1908 to the last three harpsichords, were leathered, and the 16′ was stacked on top of the 8′ bridge by using overspun strings. The instrument case was not extended to accommodate the deep register; in fact, it was made (like that of earlier Dolmetsch instruments) shorter than the Taskin prototype, with an incongruously Germanic or piano-like double-curved bentside, and, for good measure, with a heavy timber under the soundboard which actually rendered its last 23 cm ineffectual.

From Boston, Dolmetsch moved to Paris, where he continued his work at the piano manufacturers Gaveau, who had not previously made early keyboard instruments. The models produced were essentially similar to the Chickering instruments. The heavy timber member under the soundboard was abandoned, but the case was shortened still further. The 16′ register was now a standard feature of the larger Dolmetsch harpsichords. In the spring of 1914 Dolmetsch returned to England, and by 1918 he had established his workshop in Haslemere, Surrey, where his successors have maintained their factory. Gaveau continued to build harpsichords and related instruments (from Dolmetsch's plans) until the economic crisis of the 1930s.

The Pleyel firm introduced a smaller version of its 'Landowska' model concert harpsichord in 1927, still iron-framed but without a 16′ stop and descending only to *A′* instead of *F′*; the pedal action was negative as in the large model. In 1925 Dolmetsch implemented a new conception in harpsichord actions, a mechanism intended to avoid the accessory noises and jangle that can mark the passage of the plectrum past the string on the jack's return to its original position. Regulation of the new action, however, was difficult to attain and maintain; and though the action did afford the possibility of fitting a damper pedal, this was insufficient to redeem it and it was eventually discarded. A device fitted to the upper manual allowing for a kind of clavichord-like *Bebung* was a feature of Dolmetsch harpsichords for some years afterwards. A compound metal frame of wrought iron and steel welded together was introduced by Dolmetsch in 1930 but given up a few years later as it did not bring about the desired increase in stability of tuning. Modernization of the instrument was attempted by other makers as well. Karl Maendler in Munich, for instance, worked for years to develop a harpsichord with an action that would admit touch dynamics. The resulting instrument, dubbed the 'Bachklavier', was introduced with some success by the German harpsichordist Julia Menz, but it failed to survive (see Wörsching, 1946, pp.36f). Maendler's addition of a damper pedal, which raised the dampers of the lower-manual jacks only, was longer-lived. About 1933, again in response to the wishes of organists, Ammer Brothers (Eisenberg) began producing pedal keyboards with independent sets of strings and jacks which could be placed under a conventional harpsichord. Other builders, such as Neupert and Maendler-Schramm in Germany, and Alec Hodsdon in England, began a similar production shortly after.

Despite unfavourable economic conditions, professional harpsichord building in the USA, which had been suspended since the departure of Dolmetsch for Paris in 1910, was resumed in 1931 when John Challis returned to his homeland after four years at Haslemere, where he was the first Dolmetsch Foundation scholar-craftsman. In the earliest Challis harpsichords framing was wholly of wood and no adjusting screws were added to the traditional wooden jacks. But subsequent instruments reflected Challis's ingenuity in adapting the latest synthetic materials and technological advances. In his last years he achieved his aim of creating a harpsichord that would be at least as stable in the rigorous North American climate as were indigenous pianos. His late instruments were constructed wholly of metal, including the soundboard, with wood veneers used only as a decorative covering on keyboards and casework. While the tonal quality of Challis instruments – very little influenced by the sound of the early harpsichord – was not to everyone's taste, his craftsmanship was universally admired. Two pedal harpsichords built for organist clients represent the summit of his achievement. The disposition of the more elaborate of the pair set a record for sheer complexity – pedals: 16′, 8′, 4′, 2′; lower manual: 16′, 8′, 4′; upper manual: 8′, 4′; plus the usual buff stops and manual coupler.

In 1935 Thomas Goff, a London barrister, set up a workshop to build instruments to the designs of Herbert Lambert, which were influenced by both the later Dolmetsch and the modern German harpsichords. Only 14 Goff harpsichords were produced, disposed like the large Pleyel model, with metal frame and heavy case as well as heavy stringing and plectra (on later instruments both of leather and quill). They were widely used as concert instruments in the years immediately after World War II. Robert Goble, after 12 years in the Dolmetsch workshop, set up on his own in 1937, but undertook large two-manual instruments only a decade later. These sturdy wood-framed harpsichords in the modern tradition offered the resources of the Pleyel disposition but with greater volume of sound and stability of tuning and regulation.

(iii) Since 1940

Harpsichord making suffered extensively from the havoc wrought by World War II. Talented younger builders died, including Rudolph Dolmetsch, the elder son of Arnold. Maendler's workshop and others were destroyed by bombing and never regained the momentum of their pre-war years. After 1945 such surviving shops as Neupert and Pleyel resumed production much as it had been in 1939. Many renowned modern makers began learning their craft as apprentices during the postwar years: Konrad Sassmann, Kurt Wittmayer and John Feldberg at the Neupert workshop, Frank Hubbard at the Dolmetsch shop, and William Dowd and Frank Rutkowski at Challis's. Hubbard also worked briefly in London with Hugh Gough, who was influenced by Dolmetsch and who had built early keyboard instruments from 1946. Gough made relatively few harpsichords, but these were remarkable at the time for their closer resemblance to historical instruments than any modern ones since the Dolmetsch–Chickering models. After moving to the USA in 1959, however, Gough devoted himself exclusively to other types of instrument.

In 1949 Hubbard and Dowd established their joint workshop

in Boston, Massachusetts, the first in modern times dedicated to the construction of harpsichords according to historical principles. Their collaboration continued until 1958, and, in the words of Ralph Kirkpatrick, 'accomplished the major revolution of this century in harpsichord building . . . a return to seventeenth- and eighteenth-century traditions and principles of construction that had hitherto been practiced only in isolated instances'. From this point on, players were faced with a fundamental choice between the modern harpsichord as it had evolved since the beginning of the revival, and reconstructions of historical instruments.

Working independently, Martin Skowroneck of Bremen completed his first harpsichord built on historical lines in 1953. But in Germany it was specially difficult for the traditional type of instrument to gain a foothold. In no other country had the modern type of harpsichord become so firmly established. Every concert hall and radio station had acquired or had ready access to a modern instrument, invariably a large two-manual harpsichord with the spurious 'Bach' disposition. Conservatory teaching was based on this standard concert model. Performers and public alike had grown used to it, and even its appearance – because this instrument was exported round the world to an extent unparalleled by harpsichords of any other country – was a part of musical life. Though Skowroneck's work was followed in a few years by that of other historically orientated makers, such as Rainer Schütze and Klaus Ahrend, the modern instrument continued to dominate the concert stage in Germany into the 1970s. In the USA, on the other hand, the use of traditional harpsichords became widespread, the modern instrument being used almost exclusively for 20th-century music, at least by the younger generation of performers.

With the shift away from the modern harpsichord to the historical instrument, performing style has also been greatly reformed, with far less emphasis being placed on registration changes than formerly. Earlier types of harpsichord, such as models after Ruckers and the older Italian school, are coming into wider use for specialized purposes, although the large 18th-century double harpsichord has tended to assume the central role formerly occupied by the modern concert instrument. The influence of builders active in the restoration of antique harpsichords has contributed to a greater awareness of the special qualities of the best historical instruments. The new generation

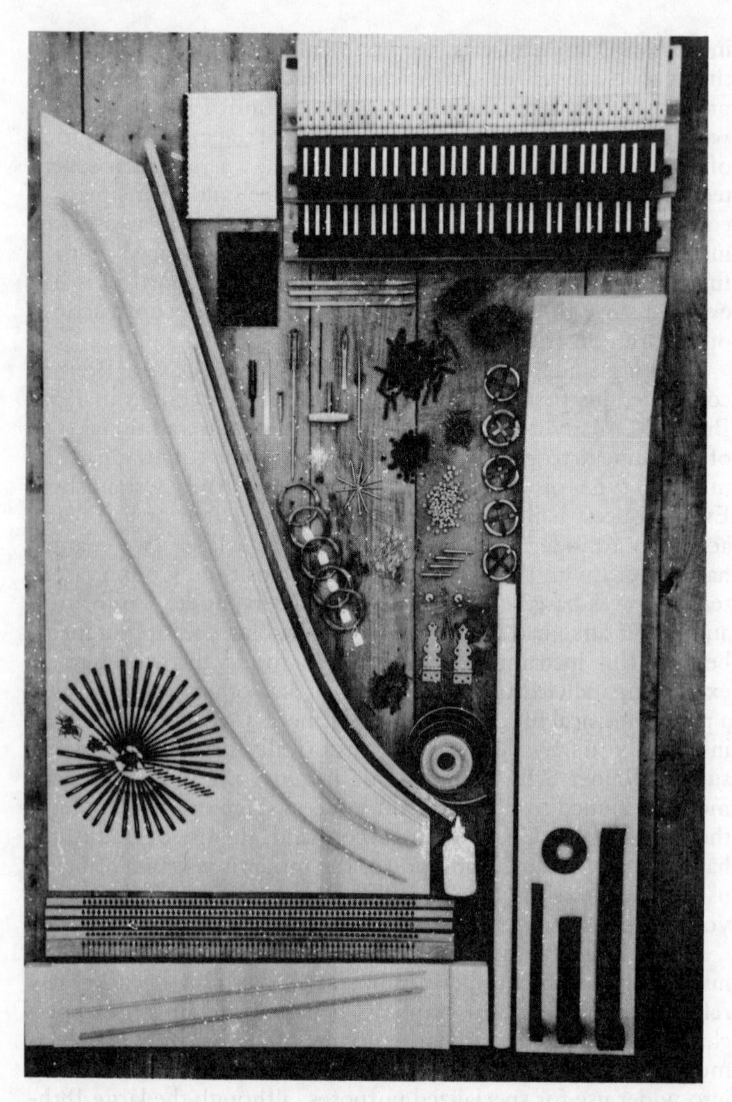

24. Complete harpsichord kit by Frank Hubbard, Boston: a replica of the harpsichord by Pascal Taskin (1769) shown in fig.17

of harpsichord makers, without significant exception, are concentrating on the historical instrument. A certain share of the credit for the growing interest in harpsichord making and playing in recent years is due to the introduction of instruments in kit form. This was pioneered by W. J. Zuckermann in 1960 with a simplified, modern type of instrument, and shortly thereafter reproductions of historical instruments in kit form were introduced by Frank Hubbard (see fig.24). In recent times the higher-quality kits have come to offer potentially excellent harpsichords of quite authentic construction and materials. It remains to be seen whether contemporary composers, who have generally favoured the modern instrument (and most often prescribed specific registration changes possible only on the pedal harpsichord with all its resources), will now accept the limitations of the classic instrument in this respect. A few composers have accepted commissions to write works for the hand-stop instrument with a classic disposition. But whatever contemporary composers may prefer, it is clear that most performers of early music have now opted for the harpsichord in its traditional form.

CHAPTER TWO

The Virginal and the Spinet

1. VIRGINAL

The virginal is a smaller type of harpsichord, usually with only one set of strings and jacks and invariably with only one keyboard. The precise application of the term 'virginal' (French and Italian *virginale*, German *Virginal*) is much debated, partly because of its use in England to denote all quilled keyboard instruments well into the 17th century. Although some writers still reserve the term for rectangular instruments, present usage generally applies it to instruments whose strings run at right angles to the keys, rather than parallel with them as in a harpsichord or at an oblique angle as in a spinet. A distinction based on the supposed uniqueness of a virginal's having two bridges resting on free soundboard is no longer tenable, since some virginals have one bridge deadened by a massive bar underneath the soundboard and since the wrest plank bridge, as well as the soundboard bridge, on some harpsichords rests on free soundboard owing to the hollowing-out of the wrest plank beneath it. However, 'spinet' is often applied to polygonal Italian instruments (in English and in German) because of the similarity to the Italian word 'spinetta' (see §2 below). In 16th-century Italy such an instrument was called an 'arpicordo' (see Chapter Four, §2). The term 'pair of virginals', to be found in early literature derived from organ terminology, denotes a single instrument.

The derivation of the term remains in dispute, the association with the Latin *virga* ('rod') being unproved and that with Elizabeth I ('the virgin queen') being without foundation. The term probably derives in some way, however, from the instrument's association with female performers – Marcuse suggested that this results from a confusion between 'timbrel' (a frame

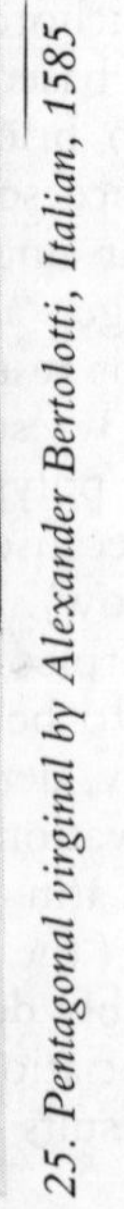

25. Pentagonal virginal by Alexander Bertolotti, Italian, 1585

drum played by women since biblical times) and the 'cymbel' in such terms as 'cembalo', 'clavicymbel' etc – or from its tone, which some theorists likened to a young girl's voice (*vox virginalis*). Other terms for the instrument include *épinette* (French), *Instrument* (German), and *spinetta* and *spinettina* (Italian).

In contrast to those of a spinet, the long bass strings of a virginal are at the front, making it possible to build the instrument in a wide variety of shapes, from squat rectangles to more or less graceful polygons, depending on whether the keyboard is inset or projecting. The rectangular form would appear to have been the earliest. It is cited in the manuscript treatise of Paulus Paulirinus of Prague (*c*1460), who described the virginal as 'an instrument having the shape of a clavichord and metal strings making the sound of a harpsichord'. This form was also the one known to Virdung, who showed a small rectangular instrument with a projecting keyboard having a range of just over three octaves (*F* to *g''*, lacking *F*♯) in his *Musica getutscht* (1511). Non-rectangular instruments appear in early 16th-century Italian representations, notably intarsias in the Vatican and Genoa Cathedral, as well as Giorgione's well-known *Concert* in the Pitti Palace, Florence. All of these date from shortly after 1510 and some show for the first time the thin-cased construction now considered as typically Italian (see fig.25). Giorgione's painting and the Genoa intarsia depict polygonal instruments, and the Vatican intarsia shows an elegant harp-shaped one, suggesting that Italian virginals were made in an even wider variety of forms than may be seen from the surviving examples dating from before the standardization of the basic polygonal and rectangular designs.

In a typical virginal the jacks are placed along a line running from the front of the instrument at the left to the back of the instrument at the right. The key-levers are correspondingly quite short in the bass and quite long in the treble, giving the keyboard of the virginal a characteristic touch, and, in some cases, making it difficult to play notes in the extreme bass easily and quickly. The jacks (one for each key) are arranged in pairs and pluck in opposite directions, so that the pairs of jacks are separated by closely spaced pairs of strings. In Italian instruments, each jack generally has its own rectangular mortise through the soundboard and a massive register beneath, whereas in north European virginals a single long mortise in the sound-

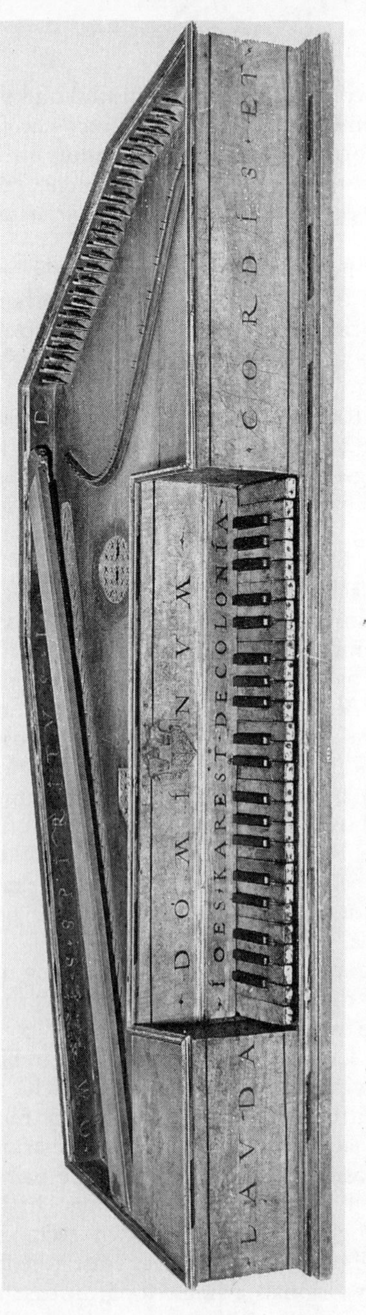

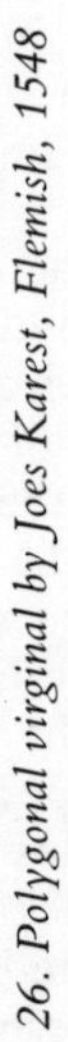

26. *Polygonal virginal by Joes Karest, Flemish, 1548*

board (together with another in a thin guide placed above the keys) serves both jacks of a pair.

Although virginals were made in more than one shape, rectangular or polygonal, the internal bracing is similar for the different types. There is a brace extending from the front to the back of the case at each side of the keyboard; this may be supplemented by knees (triangular blocks) bracing the sides to the bottom and by corner blocks. A liner around the inside of the case supports the soundboard and carries the hitch-pins; the wrest-pins (tuning pins) are held by a larger piece of hardwood. In rectangular instruments a separate diagonal hitch-pin rail and wrest plank are provided. The back corners of rectangular Italian virginals were sometimes cut off as shown by the 1593 Celestini virginal (Donaldson Collection, Royal College of Music, London). A few employ the false inner-outer construction in which a thick softwood case is fitted with cypress veneer and half-mouldings to make it appear as if a cypress instrument were in an outer case. By far the most common Italian construction used thin (3–5 mm) cypress case sides with mouldings on the top and bottom edges, as in Italian harpsichords.

The case joints of Italian virginals are mitred, as might be expected considering the thinness of their wood; accordingly, it is noteworthy that the corners of the 1548 Karest virginal (fig.26, p.114) are dovetailed, even though the wood is scarcely thicker. Otherwise, the structure of the instrument is hardly different from that of an Italian example, except for the replacement of the solid Italian jack register by a complete counter-soundboard mortised for the jacks and serving as a lower guide (this feature is also found in a number of German and English instruments, and dovetailing of the case joints in German harpsichords and clavichords persisted into the 18th century).

Later Flemish virginals have thick cases, like those of Flemish harpsichords, and are assembled before the bottom is put on rather than being built from the bottom up. As with the Italian instruments, however, the two principal braces run from the front of the instrument to the back at the ends of the keyboard. The decoration of these instruments normally corresponds to that of the Flemish harpsichords, either plain paint or marbling outside and block-printed papers inside, except for the inside of the lid, which may have a painted landscape on it. A few examples are painted with arabesques instead of being papered, but this must have been relatively uncommon. The inclusion of

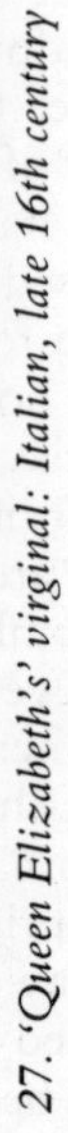

27. 'Queen Elizabeth's' virginal: Italian, late 16th century

a Latin motto either on the inside of the lid, on the jackrail, or around the inside of the case above the soundboard was quite common, and this is also seen in German instruments.

English virginals, although constructed in much the same way as Flemish ones, differ in being made of oak rather than of the lime or poplar used in the Low Countries. They are also distinguished by the retention of the vaulted lid, which seems to have gone out of use elsewhere early in the 17th century. The dark oak of the outside of these instruments is in striking contrast to the gilt paper and brilliant painting discovered when the lid is lifted and the hinged front board lowered. English virginals tend to have a wide range: from F' or G' in the bass to d''', with a chromatic bass. It is evident from the music of around 1600 that notes lower than *C*, and an $F\sharp$ or $G\sharp$ are required, indicating that English virginals were built to meet these requirements exactly. Flemish and Italian virginals usually ended on a *C/E* short octave. A well-known instrument, the so-called 'Queen Elizabeth's Virginal' (in the Victoria and Albert Museum, London; see fig.27, p.116), is Italian, probably Venetian. Its original keyboard had a *C/E* to f''' range, but a *C* to c''',d''' (i.e. no $c\sharp'''$) keyboard was later installed, presumably in order to alleviate the difficulty of the non-chromatic bass. In turn, this keyboard was later modified to the present G'/B' to c''' range (see Wraight, 1986).

Numerous Italian 16th-century virginals have survived and show that often considerable trouble was taken to provide elaborate decoration: the 1577 Rossi instrument (Victoria and Albert Museum, London), with inset precious and semi-precious stones is a fine piece of work (see cover illustration). More typical was a plain cypress case with fine mouldings, but sometimes the casework was set with intarsia work in contrasting coloured woods, as in the 1523 Francesci de Portalupis virginal (Musée Instrumental, Paris Conservatoire). All of these virginals have a rose, usually made of three or four layers of wood veneer, pierced in intricate gothic or geometric designs. Sometimes carved brackets were set either side of the keyboard. Ivory was sometimes used for the natural-key covers, but boxwood was most commonly used, with ebony-topped sharps. Most instruments by Venetian makers had projecting keyboards, and most virginals from the Milan–Brescia area had partly recessed keyboards, but exceptions to these traditions are found in both regions. There is no overwhelming acoustical advantage in one system or the other. Almost all 16th-century virginals have a *C/E* to *f'''* compass and later 17th-century examples only *C/E* to *c'''*. Early 16th-century instruments were made with an *F,G,A* to *f'''* compass, but in the few instruments that survive the keyboard has since been modified. Some 17th-century virginals had split keys for *d♯/e♭* and *g♯/a♭*.

Modifications, which have obscured the history of the Italian harpsichord, were not undertaken often for the virginals. It was relatively difficult to replace keyboards with those from other instruments since the pairwise jackslide tends only to match the keyboards for which it was made. Nevertheless, some virginals had their scales changed and keyboards altered to keep them abreast of changing musical requirements.

Considerable discussion has been devoted to the question of the pitch of Italian virginals (see also Chapter One, §2(i)): it has been argued that the long scales (usually corresponding to *C/E* to *f'''* compasses) were intended for low pitches. Most 16th-century Italian virginals have scales between 30.5 and 35 cm at *c''* (with only a few being shorter than this). These long-scaled instruments were, however, designed to be strung in iron wire, which means that this long scale came to a normal pitch (i.e. for 35 cm about a tone below *a'* = 440 Hz).

Although most Italian virginals were designed for iron wire, some were quite clearly intended for brass wire at this same pitch. Several 17th-century rectangular instruments, by Guarracino, are examples of the latter. They have the tuning pins on the left-hand side and only one bridge on free soundboard, with the result that the sound is much brighter than that of most virginals and closely resembles that of a bentside spinet. These virginals by Guarracino appear to have been the continuation of a Neapolitan tradition: a similar instrument was made by Alessandro Fabri in 1598 (Tagliavini and Van de Meer, 1986). Since instruments of this design are unknown further north of the Italian peninsular, the speculation is permissible that this design might have been influenced by Spanish virginals. Virtually nothing is known of the latter, but Naples was under Spanish administration at that time so the introduction of Spanish instruments would have been possible.

Flemish makers produced the muselar (see fig.30, p.124) and spinett (see fig.29, p.122), in which the difference in sound results from the different plucking points. All Italian makers on the other hand adopted similar scalings and plucking points, thereby giving a fairly uniform character to their instruments. The sound of an Italian virginal is usually louder than that of an Italian harpsichord, since the virginal has two bridges on free soundboard. However, as with all keyboard instruments with a plucking action, it is possible to vary the volume considerably by voicing the plectra.

Some instrument makers had a better reputation than others: Dominicus Pisaurensis (Domenico da Pesaro), who was praised by Zarlino (1558), made several virginals that have survived. No virginals by Vito Trasuntino have survived (an instrument bearing his name has a fake inscription), but it is recorded that he made excellent instruments. Alessandro Trasuntino came to an arrangement with Pietro Aretino to make him an arpicordo, for which he was to receive a painting in exchange.

A few virginals were made with two sets of 8′ strings although this was not common. Two surviving instruments were made by Donatus Undeus (one, 1623, in the Brussels Museum of Musical Instruments, and one undated, but attributed by Wraight to him, now in the Benton Fletcher Collection, Fenton House, Hampstead, London). Another rectangular virginal, made by Celestini in 1610 (Brussels Museum of Musical Instruments) does not have the usual pairwise arrangement of jacks (i.e. pairs of jacks separated by pairs of strings), instead each string is separated from the next string by one jack. On the nameboard this instrument is described as an 'arcispineta'.

Although the tonal resources of the virginal (which has a single set of strings and is seldom equipped with any means of changing timbre) are more limited than the harpsichord's, it none the less occupies a crucial position both in musical life and in the development of quilled instruments in general in the 16th and 17th centuries. Harpsichords certainly existed throughout this period both north and south of the Alps, but they are more rarely represented in paintings, drawings etc than virginals, and it must be concluded that they were much less common than the smaller, simpler and cheaper instrument. In addition, the sound of virginals is excellently suited to most of the keyboard literature of the period. That the virginal occupied a central position in Flemish instrument building is well illustrated by the fact that in the rules for the admission of keyboard instrument makers to the Antwerp Guild of St Luke (drafted in 1557) the piece of work to be submitted by a candidate was specifically

MVSICA·DISPARIVM·DVLCIS·CONCORDIA·VOCVM
PELLO·LEVO·PLACO·TRISTIA·CORDA·DEOS
ORGANO
LAVDATE

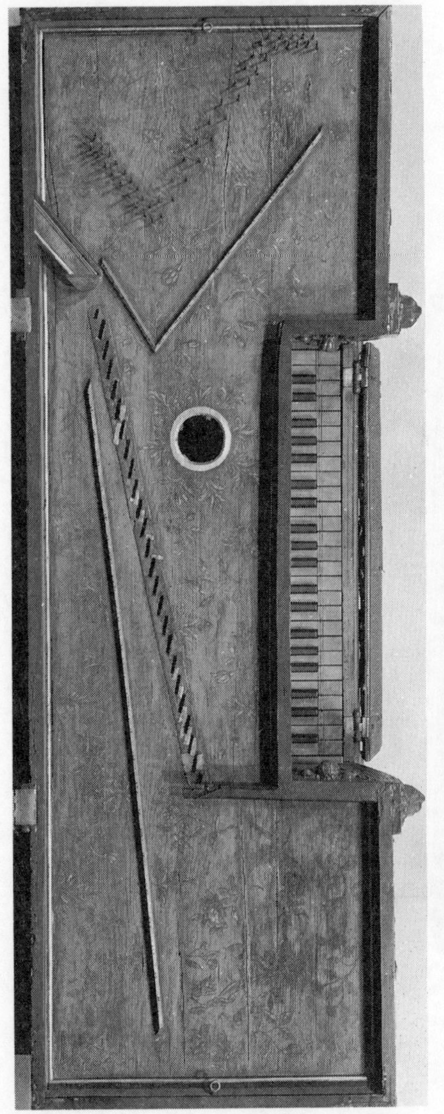

28. Virginal with centrally placed keyboard, Flemish, 1568, general and plan views

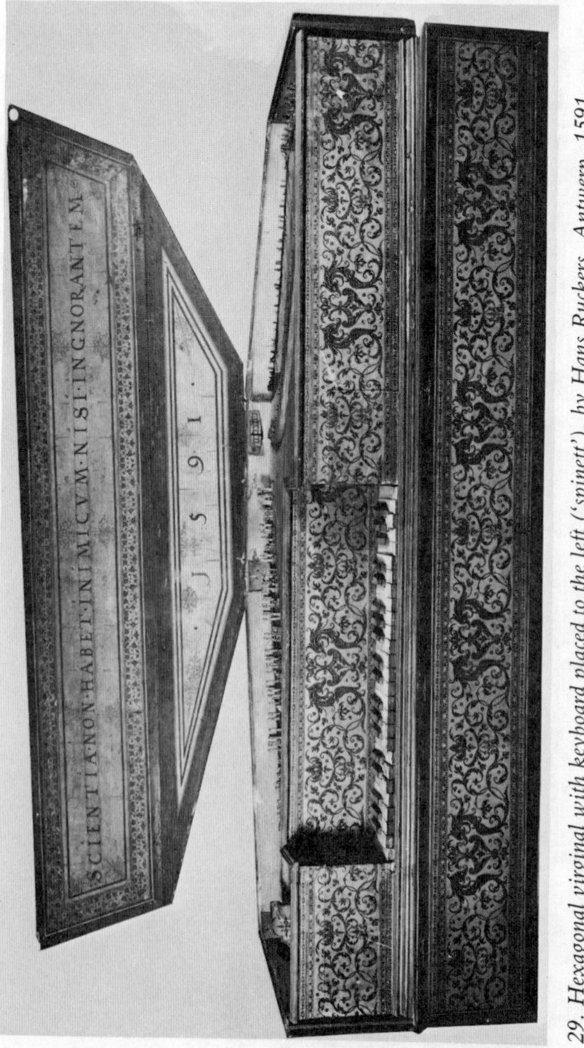

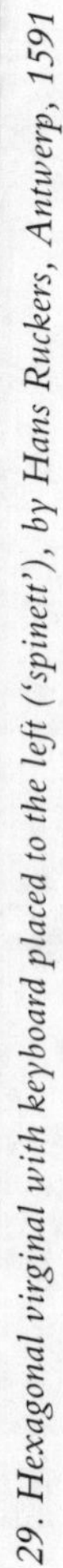

29. Hexagonal virginal with keyboard placed to the left ('spinett'), by Hans Ruckers, Antwerp, 1591

designated as 'a rectangular or polygonal virginal' ('een viercante of gehoecte clavisymbale') and no mention of harpsichords is made at all. Furthermore, O'Brien (1974) has shown that the Ruckers family made many more virginals than harpsichords. (See also Chapter One, §3(i).)

The earliest surviving Flemish virginal (see fig.26), like those depicted in Flemish paintings from before 1565, is thin-cased and polygonal. These instruments thus bear a superficial resemblance to their Italian counterparts, the most obvious difference being that their keyboards are entirely recessed rather than wholly or partly projecting; however, some Italian virginals of false inner–outer design have completely recessed keyboards. The appearance of a Flemish polygonal instrument seems somewhat heavier than the graceful Italian design. Ripin (1971) has speculated that the Flemish polygonal design was transmitted to the Low Countries by way of Germany. It is known that some of the earliest 15th-century makers working in the Low Countries came from Cologne. Of Hans van Cuelen (Hans from Cologne) little is known, but Joes Karest headed the instrument makers' petition to the Guild of St Luke in 1557 as they sought to be admitted as instrument makers and not grouped with the painters. A German harpsichord of 1537 made by Hans Müller in Leipzig shows striking resemblances to thin-cased Italian harpsichords. However, the answer as to where these harpsichord or virginal designs orginated probably lies in the 15th century, and about this period so little is known in detail that answers can be no more than speculative. As is often the case with inventions, similar designs may have been developed simultaneously in different parts of Europe.

As noted above, however, the guild regulations make it clear that rectangular as well as polygonal instruments were being made in Antwerp in the 1550s, and presumably these instruments were also thin-cased. Although no example from this period survives, the earliest Flemish depiction of a rectangular virginal (an engraving by Cornelis Cort printed in 1565, based on a painting by Frans Floris from ten years earlier, now lost) shows an instrument without a lid, suggesting that it was thin-cased and intended to be kept in a stout outer case like an Italian instrument. By the end of the 1560s, however, it would seem that a thick-cased rectangular instrument very much like those now considered typically Flemish had come into being, since a painting by Michael Coxcie purchased by Philip II of Spain in

30. Double virginal with keyboard placed to the right ('muselar'), by Hans Ruckers, Antwerp, 1581; the octave instrument, or 'child', is not shown, but would be placed over the jacks behind the keyboard (see fig.31)

1569 clearly shows an instrument of this kind, which seems immediately to have superseded the thin-cased types in the Low Countries.

At the end of the 16th century, three types of virginal were being made in Flanders: one with the keyboard centred in one of the longer sides (fig.28), one with it placed off-centre to the left (called 'spinett' by Claas Douwes; see fig.29), and one with it placed off-centre to the right ('muselar'; fig.30). The centre-keyboard design in normal-pitch virginals is known only from one instrument, that of the Duke of Cleves (fig.28) and it is impossible to say whether this was an isolated example, or whether it was a design which was later forgotten. Octave instruments retain their central placement of the keyboard, although this is virtually obligatory because of their small size. The difference in the placement of the keyboard is important since it determines the placement of the jacks in relation to the two bridges. With the keyboard placed to the left, the jacks run in a line close to the left-hand bridge; where the keyboard is centrally placed, the jacks in the bass are still well away from the centre of the string. This results in the point at which the jacks pluck the strings gradually shifting from near the centre of the strings in the treble to a position far removed from the centre in the bass. Because of this shift in the plucking-point, the timbre of these virginals gradually changes from flute-like in the treble to reedy in the bass, being similar in this respect to the timbre of a harpsichord. Virginals with their keyboard at the right, on the contrary, have their strings plucked at a point near the centre for virtually their entire range, producing a powerful, flute-like tone that varies little from treble to bass. Both spinetten and muselars were made in a variety of sizes, the smaller ones presumably tuned to higher pitches and the smallest ones clearly tuned an octave above the largest ones. Among the surviving instruments, muselars are more numerous in the full-size examples, spinetten in the smaller sizes.

The earliest surviving example of a muselar is by Hans Ruckers (*d* 1598) and is dated 1581 (fig.30); it is entirely possible that Ruckers invented the muselar design, even though it must be emphasized that the other characteristics now associated with Ruckers's work are to be found in virginals made by the preceding generation of makers, notably Hans Bos and Marten van der Biest. Muselars always have both their bridges resting on free soundboard in contrast to the spinetten and the polygonal

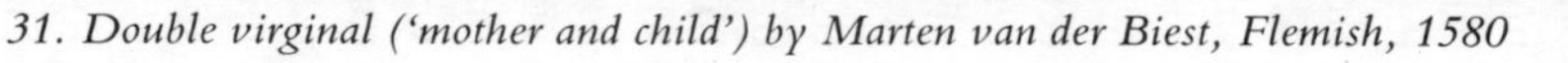

31. *Double virginal ('mother and child') by Marten van der Biest, Flemish, 1580*

instruments, which often appear to have had the bridge at the left deadened. Ruckers's practice in this regard was not consistent; the smaller spinetten all seem to have the left-hand bridge deadened, but some of the full-size examples have both bridges resting on free soundboard. Muselars have the further distinction of being the only virginals normally provided with any means of changing timbre. A substantial number of the surviving examples have an *arpichordum* stop, which may be engaged to produce a buzzing sound in the tenor and bass to contrast with the clear flute-like sound of the alto and treble.

The culmination of the Flemish virginal makers' art was the double virginal called 'mother and child' by Joos Verschuere-Reynvaan (*Muzijkaal kunst-woordenboek*, 2/1795), a usage apparently sanctioned by the makers, since the original Ruckers numbers on surviving examples include an 'M' and a 'k' for *moeder* and *kind* on the large and small instruments respectively. These instruments consisted of a virginal of normal size (the 'mother') with a compartment next to its off-centre keyboard in which a removable octave instrument (the 'child') was housed (fig.31). The octave instrument was designed to be coupled directly to the larger one. An oblique slot was cut in the bottom of the octave instrument, and if the jackrail of the 'mother' was removed and the 'child' was put on top of it in place of the jackrail, the larger instrument's jacks reached through the slot to touch the underside of the octave instrument's key-levers. Thus when a key of the 'mother' instrument was depressed its jack pushed upwards on the back end of the corresponding key of the 'child', causing the octave instrument to sound at the same time as the larger one. In addition, the octave instrument could be played separately, either when in place on top of the larger instrument or entirely removed from it. Instruments of this kind appear to have been imitated in Germany and Austria, and an Innsbruck inventory made in 1665 mentions a virginal in which two smaller ones were contained, 'all three of which could be placed on one another and sounded together'.

Praetorius (*Syntagma musicum*, ii), who also mentioned placing virginals on top of one another, depicted a rectangular instrument differing in several respects from the developed Flemish design, rather resembling pre-Ruckers examples in having a centred keyboard and a jackrail supported at the left by an arm extending from the bass end of the key-well. The instrument shown by Praetorius also has a larger range than the regular

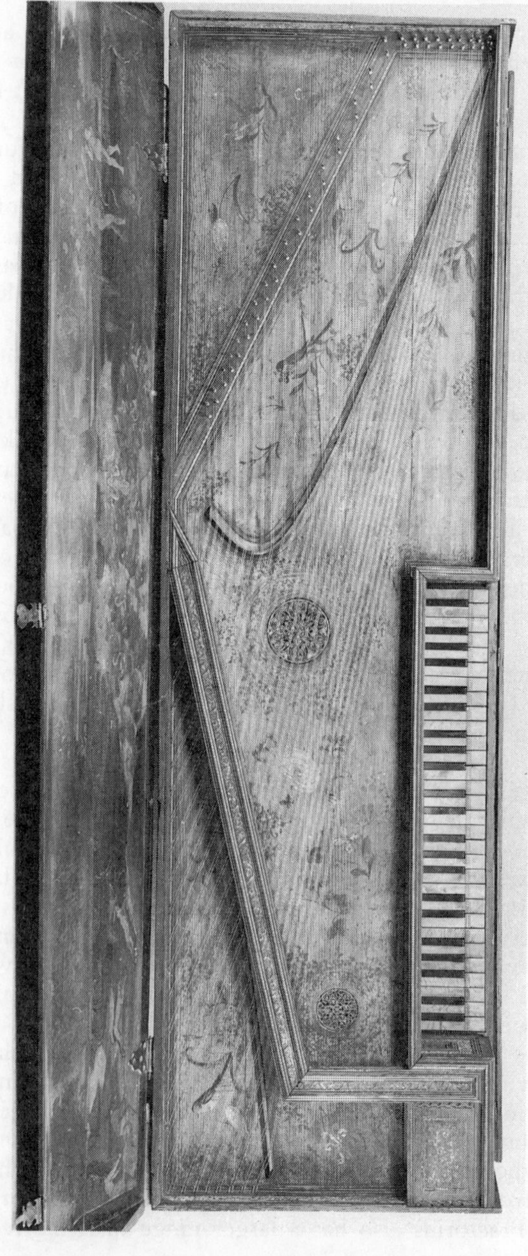

32. Plan view of rectangular virginal by John Loosemore, Exeter, 1655; the placement of the bridge makes the sounding lengths of the strings shorter than in fig.30

Flemish virginal compass of four octaves with a bass short octave, *C/E*, to *c'''*, its keyboard extending to *d'''*, with divided accidentals in the lowest octave to provide the *F♯* and *G♯* not ordinarily available with the *C/E* short octave, as well as having the *E♭* keys divided in the remaining octaves to sound *D♯*, thereby extending the range of available keys to include E major and E minor. As with virtually all north European virginals, a tool box is set in the space to the left of the keyboard.

Although north European polygonal virginals were normally housed in outer cases like their Italian counterparts, this does not seem always to have been true of the rectangular instruments. Praetorius did not show a lid for the instrument he illustrated (presumably to save space), although the presence of a hinged front board to cover the keys strongly suggests that the instrument had one; it would be of great interest to know whether such a lid would have been vaulted like those shown in French illustrations of the 1580s, on the title-page of *Parthenia* (London, 1612–13; see fig.55, p.209) and found on all surviving English virginals, which range in date from 1641 to 1679. Despite the fact that the English instruments seem to be patterned on Flemish spinetten, the vaulted lid, the method of supporting the left end of the jackrail and the shorter scaling of English virginals (fig.32) suggest that they actually derive from the same non-Flemish tradition represented by Praetorius's illustration and the surviving 17th-century harpsichords and virginals from Germany and France – a tradition also represented by the polygonal Flemish virginals from the mid-16th century.

Outside Italy the making of rectangular virginals seems to have come to an end by the close of the 17th century, these instruments having been replaced by bentside spinets. Since only one bridge of a spinet is on free soundboard, the sound of a spinet resembles that of a harpsichord and it may be for this reason that taste changed in favour of the spinet and against the virginal. Although the spinet is typically somewhat smaller than a rectangular virginal, the scalings employed in both instruments are similar. Polygonal virginals and spinets usually have less soundboard area in the low tenor and bass than rectangular virginals; this can impart a clearer, reedier character to the sound, but as in all keyboard instruments, it is the skill of the maker that ultimately influences the quality of the sound. Some virginals were made in the early 19th century in Italy: they resemble square pianos in appearance, but contain a plucking

33. 'A Man and a Woman Seated at a Virginal': painting by Gabriel Metsu (1629–67)

mechanism. The last dated of these (1839, now in the Musikinstrumenten-Museum, Karl-Marx-Universität, Leipzig) is by Alessandro Riva of Bergamo.

The importance of the virginal has been exaggerated by some, who, unaware that the term was used for all plucked keyboard instruments in England, assume that the music of Byrd, Bull, Gibbons, Tomkins and others was intended specifically for these single-strung instruments. However, as is made clear by the title-page illustration of *Parthenia In-violata or Mayden-musicke for the Virginalls and Bass-viol* (*c*1624), which shows a harpsichord rather than the rectangular instrument of *Parthenia*, the music of the 'virginalist composers' was not intended to be restricted to the virginal. Other writers have relegated the instrument to a status rather lower than that of the modern upright piano. The proper assessment of the virginal lies between these extremes. Despite the limitations imposed by a single register, virginals are useful and remarkably versatile instruments with special qualities of their own, on which virtually the entire literature of their period can be played with considerable success. Thus, although virginals were often used merely as practice instruments and for domestic music-making, they should not, like the spinet, be thought of essentially as substitutes for the harpsichord.

2. SPINET

The spinet is a small keyboard instrument with a plucking mechanism as in the harpsichord (see Chapter One, §1). It usually has a single keyboard and one set of jacks and strings. The word 'spinet' may derive from the Latin *spina*, 'thorn', referring to the small quill which plucks the strings. Banchieri (2/1611) suggested that this type of instrument was invented by Giovanni Spinetti at the turn of the 16th century, but that is unlikely. As with the virginal, there has been some difficulty in defining the characteristics that an instrument must have in order to be called a spinet. In the preferred current usage, 'spinet' refers to an instrument whose strings run diagonally from left to right with respect to the keyboard (in the harpsichord the strings run directly away from the keyboard, and in the virginal transversely). Another usage reserves 'virginal' for rectangular

34. Bentside spinet by John Harrison, London, 1757

instruments and 'spinet' for other forms. Thus there is often some confusion as to whether an instrument should be called 'spinet' or 'virginal'. The difficulty lies in the fact that although virtually the same word has been used in English, French, German, Italian and Dutch, the instruments designated are not identical. Thus, the English 'spinet' (or 'bentside spinet') denotes the type of instrument shown in fig.34, which is called in German 'Querspinett' or 'Spinett' and is similar to the Italian bentside spinet (see fig.35, p.134). The French 'épinette' is a more general term which was used for all quilled keyboard instruments until well into the 17th century. Claas Douwes (1699) used 'spinett' to distinguish rectangular virginals which have their keyboards at the left from the centre-plucking 'muselaar' (muselar) which has its keyboard at the right. Quirinus van Blankenburg (*Elementa musica*, The Hague, 1739) implied a similar understanding of the term in naming the close-plucking lute stop of his four-register harpsichord 'spinetta'. Although some 17th-century Italian instruments were made with a bentside (fig.35) others were rectangular, such as the 1610 instrument by Celestini (in the Brussels Museum of Musical Instruments), described by the maker as 'arcispineta' on the nameboard. The term 'spinetta' was used generically in Italy for instruments which were not harpsichords, but in the 16th century 'arpicordo' (see Chapter Four, §2) was widely used for such instruments. 'Spinettina' (or 'spinettino') and 'spinetta ottavina' denote a 4′ instrument and 'spinettone' a large bentside spinet, such as the 'spinettone da orchestra' made by Cristofori and mentioned in a Medici inventory of 1700 (Gai, 1969). Another name for the *spinettone* was 'cembalo traverso'. In the USA in the 1930s the term 'spinet' was also applied to miniature upright pianos. Early rectangular pianos have been erroneously described as spinets in some auction catalogues and elsewhere.

The oblique stringing of a spinet produces a trapezoid in the smaller instruments and a wing shape in the larger ones, whose bass strings are longer than the keyboard. The longest strings of a bentside or trapeze-shaped spinet are at the back, and the tuning pins are set in a wrest plank directly over the keys instead of at the right-hand end of the case. One of the bridges over which the strings of a spinet pass is attached to the wrest plank instead of resting on a free soundboard. For this reason the sound of a spinet more closely resembles that of a harpsichord of

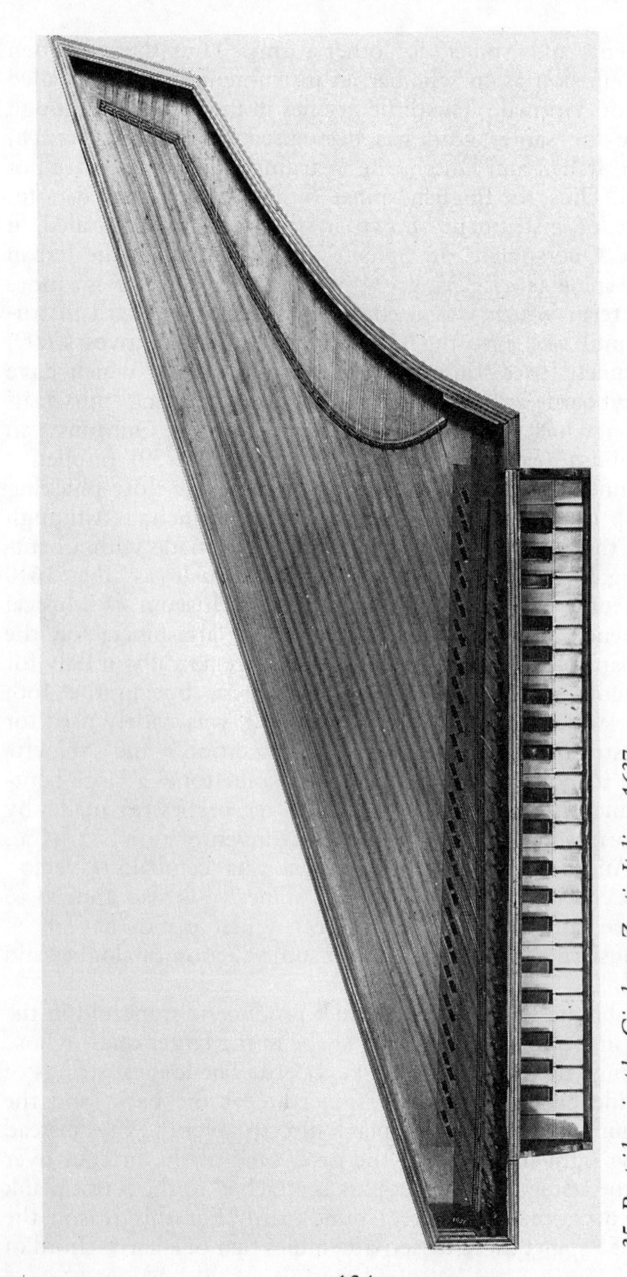

35. Bentside spinet by Girolamo Zenti, Italian, 1637

similar size than that of a virginal. However, a virginal with jacks that pluck close to the nut (e.g. the Flemish 'spinet') sounds more like a harpsichord than the centre-plucking muselar or other types of virginal. Nevertheless, the spinet has only one bridge on free soundboard, whereas most virginals have two. Comparisons of this sort always encounter the difficulty that individual instruments, because of their particular construction, may not sound like many others of the same type.

Apart from a small number of tiny rectangular instruments, made in Germany in the late 16th century and often equipped with a pin-barrel mechanism, the earliest surviving spinets are early 17th-century Italian. They have two straight sides set perpendicular to the keyboard, the left one shorter than the right. The back of the case thus slants away from the keyboard and runs parallel to the strings (see fig.36, p.137). These small instruments were not built for normal pitch, but for its octave. The design is obviously a compact one and involves compromises which are avoided in spinets of normal size. In the smaller instruments, for example, the bridge is usually made of several small pieces of wood glued on to the soundboard, since the tight curve required is difficult to bend.

The keyboard of the earliest spinets occupies virtually the entire case, leaving little room for internal structure. The sides and back of the case overlap the edges of the bottom; the wrest plank is supported by a block at each end, and these blocks are attached to the bottom and to the shorter sides of the case. The single set of jacks runs in a line in pairs, the members of which face in opposite directions, immediately behind the wrest plank (see fig.37, p.138). There is only one string per note, and no buff stop or other means of changing tone-colour.

The wing-shaped 'bentside' or 'leg-of-mutton' spinet which was to become the normal English domestic keyboard instrument in the late 17th century appears to have been invented by a widely travelled Italian, Girolamo Zenti, whom Giovanni Bontempi praised in 1695 for having created the 'most modern harpsichord . . . in the form of a nonequilateral triangle'. Bontempi went on to speak of these instruments as having two keyboards and three registers, leaving this interpretation open to doubt; however, the earliest example of a bentside spinet, dated 1637, bears Zenti's signature (fig.35). An Italian bentside spinet bearing the name Pentorisi may have been made earlier than the Zenti, but its date, 1590 or 1690, is not clear. A few other Italian bentside spinets survive, together with an even smaller number of French or German examples. The instrument had its greatest popularity in England, where it began to replace the rectangular virginal in the last decades of the 17th century. Early examples are generally made of or veneered in walnut, with ebony natural keys and solid ivory sharps. These English spinets usually have short scalings intended for brass wire at normal pitch (26–28 cm at *c''*) although some instruments were intended for iron wire at this pitch (*c''* about 32 cm). Later examples are usually veneered in mahogany and have ivory natural keys and ebony sharps (see fig.34). The surviving 18th-century American spinets are closely modelled on the English type.

The keyboard of a bentside spinet, like that of the earlier trapezoidal examples, occupies most of the case. There is usually a brace from the front of the case to the back at each end of the keyboard, and the Italian examples, as well as some 17th-century English ones, may have a few triangular knees (blocks) between the sides and bottom of the case in the unobstructed space to the right of the keyboard. Later English spinets, although retaining the projecting keyboard and single thick jack guide of Italian virginals and harpsichords, incorporate north European techniques of construction. Most 18th-century English bentside spinets employ two series of braces, one just under the soundboard and the other in the lower part of the case, a plan similar to that of north European harpsichords. There are usually only two lower braces, one at each end of the keyboard; sometimes a third brace runs transversely behind the keyboard. The wrest plank rests on a raised section of the braces at the ends of the keyboard; the bottom of the instrument is fastened to the lower edge of the braces after the construction of the case has been completed and the soundboard installed. The upper braces, usually three in number, pass from the bentside to the spine; they are attached to the lower edges of the liners which support the soundboard, and braced to the face of the liners with small triangular blocks. The many such spinets still in playable condition prove the efficiency of this simple design.

The bentside spinet is a compact instrument. Whereas the harpsichord must always be at least a foot (12 inches; 30.5 cm) longer than its longest string, the spinet need be only a few inches longer; the performer sits in front of the instrument instead of at the end. The oblique stringing of the spinet produces an instrument which is neither as wide nor as long as a

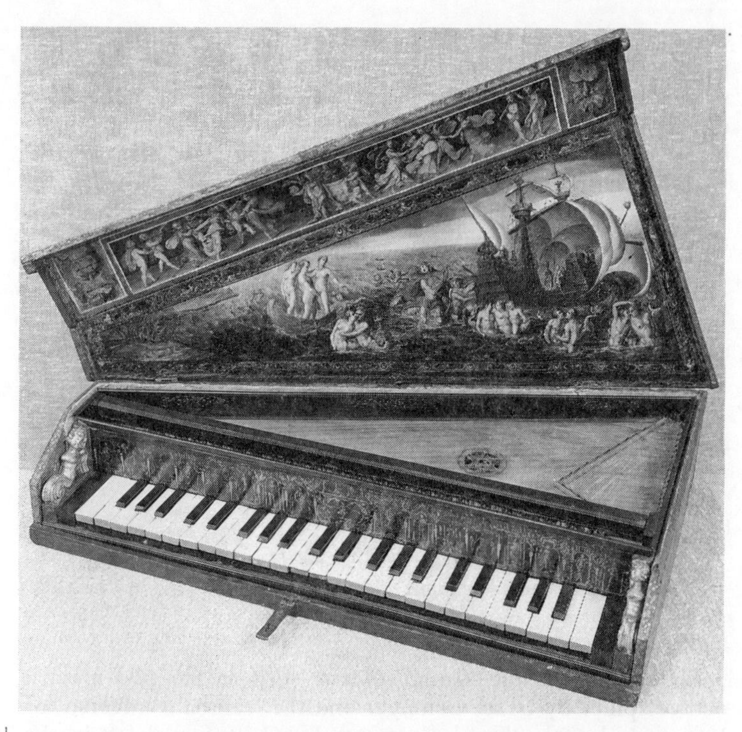

36. Trapeze-shape spinet (compass C/E to c'''), Italian, c1600

harpsichord of equal compass. It is not only the compactness of the design which leads to the small size of spinets. A short scale was usually chosen (about 26–28 cm at *c''*) in order to reduce the space requirements. Nevertheless, the ultimate size of the instrument depends on the length of the bass strings, which were shorter than in harpsichords. This has the consequence that the notes below *C* in the spinet may have a poor tone.

Neither the spinet nor the virginal is normally capable of variation in tone-colour or volume. Since the jacks are usually placed obliquely in the jack guide and face alternately in opposite directions, any movement advances half the jacks but withdraws the other half; uniform lateral movement with respect to the strings is not possible. Similarly, because both strings of each pair form part of the spinet's single register, it is not possible to employ the type of buff stop found on harpsichords. Both these problems are solved on certain modern spinets in which the strings are not arranged in pairs and all the jacks face in the same direction; such instruments can have both a buff stop and a half-hitch or 'piano' position which permits the jacks to be partly withdrawn from the strings. The 1610 'arcispineta' made by Celestini has all the jacks facing in the same direction, although there is no buff or other stop.

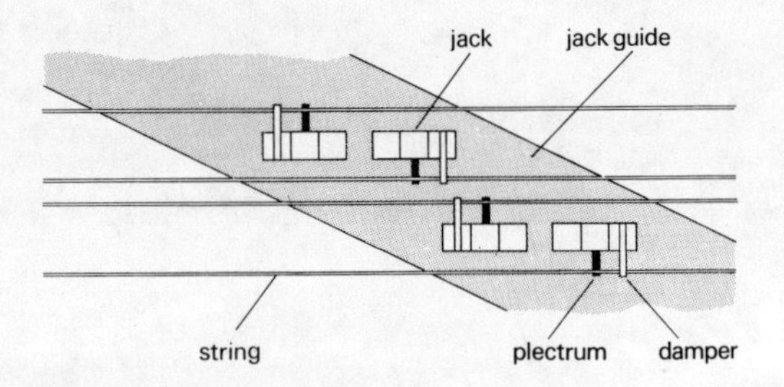

37. Diagram to show the side-by-side placement of jacks found in the spinet

In the rare double-strung spinets, such as the two-manual octave spinet by Israel Gellinger and the 'cembalo traverso' or 'spinettone' with 8′ and 4′ strings by Cristofori (both in the Musikinstrumenten-Museum, Karl-Marx-Universität, Leipzig), change in tone-colour or volume may be obtained by moving the keyboard in or out so that both or only one set of jacks will be lifted when the keys are depressed. Such elaboration is, however, exceptional and essentially foreign to the nature of the spinet, which is basically a simple, single-strung instrument.

Always a domestic keyboard instrument, the spinet cannot be said to have a repertory of its own distinct from that of the harpsichord (see Chapter Five). However, much of the music printed in such collections as *Musick's Hand-maid* (1663, 1689), *The Harpsichord Miscellany* (2 vols., *c*1763) and *The Harpsichord Master* (1697–1734) was doubtless intended for use by the amateur performer who had no larger instrument at his disposal.

CHAPTER THREE

The Clavichord

The clavichord is the simplest and at the same time the most subtle and expressive of those keyboard instruments whose sound is produced by strings rather than by pipes. It is likely that the mysterious chekker of the 14th century was in fact a clavichord (see Chapter Four, §1); in any event, it is clear from both pictures and writings that clavichords not too unlike those that are known from surviving examples were in existence in the early years of the 15th century. The clavichord was used throughout western Europe during the Renaissance and in Germany until the early 19th century, but for most of its long history was primarily valued as an instrument on which to learn, to practise and occasionally to compose. It appears as an alternative to the virginal and the harpsichord on the title-pages of a number of 16th-century keyboard collections and is similarly mentioned in some German publications of the late 17th century, but there seems to have been little or no recognition of the clavichord's special capabilities before the beginning of the 18th century and no music specifically composed for it before the mid-18th century. At that time a large body of music written particularly for the clavichord or alternatively for the clavichord and the relatively new fortepiano began to be composed in Germany, where the clavichord was still quite common. Some of this music is highly demanding, but much of it was composed for amateurs of modest capabilities and is accordingly far easier to play.

Although the clavichord is *par excellence* a solo instrument on which one plays by oneself for oneself, it was also used in the 18th century to accompany solo singing; it is specifically cited in this connection by writers of the period and on the title-pages of several song collections. (See also Chapter Five.)

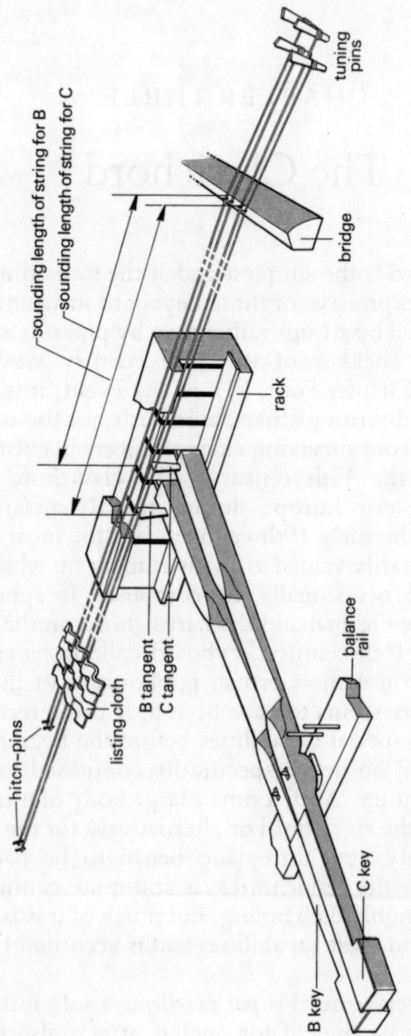

38. Mechanism of the clavichord

1. STRUCTURE AND TONE PRODUCTION

The usual shape of the clavichord is a rectangular box, with the keyboard set into or projecting from one of the longer sides. The strings pass from hitch-pins near the left end of the box across the back half of the keys and over the bridge and soundboard to the tuning pins near the right end. Each key rests on a transverse piece of wood called a balance rail, which acts as a fulcrum when the key is depressed. At the point of the fulcrum the position of the key is maintained by a pin (balance pin), which passes through a slot in the key and is driven into the balance rail below. In the back end of each key is driven a slip or blade of wood, horn or whalebone, which slides in one of a series of vertical slots cut in a piece of wood running along the inside of the back of the case. Some clavichords have a vertical guide pin set in a rail under the key end; in the key-lever is a leather-lined slot which engages the pin. This arrangement is quieter than the common one and therefore an advantage for the small volume of the clavichord. Between this slotted rack (sometimes called a diapason) and the balance rail, a brass blade called a tangent is driven down into each key; when the front of the key is depressed by the player's finger, this blade rises until its top edge strikes the pair of strings above it (at a point between the hitch-pins and the bridge), setting them into vibration. The vibrations of the section of each string between the tangent and the bridge are communicated to the soundboard, yielding a tone of small volume but great sensitivity and flexibility. The vibrations of the section of each string between the tangent and the hitch-pins are damped out by strips of cloth called 'listing' that are woven between the strings; when the key is released, causing the tangent to drop from the string, this cloth immediately damps out the vibrations of the string as a whole, instantly silencing the tone (see fig.38).

The loudness of the tone depends on the force with which the tangent strikes the strings and thus is under the direct control of the player. Moreover, as the tangent remains in contact with the strings while they are sounding, the performer can continue to influence the sound of a note after it has been struck. By increasing or decreasing the pressure on the key, the pitch of a note can be altered after it has begun, thereby producing a portamento or a vibrato (*Bebung*), which creates the illusion, within its quiet range, of swelling the tone. While striking the key quickly produces a louder sound than depressing it slowly, too much pressure lifts the strings too far, increasing their tension and distorting their pitch. The dynamic range between the instrument's all but inaudible *pianissimo* and this rather limited *fortissimo* is, however, quite significant: within it the performer can achieve the most sensitive possible control of dynamic effects.

The soft tone of the clavichord results from the small amount of force with which the tangent sets the string in motion. Two factors are involved in determining this force: the mass of the tangent (and key-lever), and the velocity with which it reaches the string. The velocity of the tangent is determined by the distance between the tangent and the string, and the acceleration applied to the tangent by pressing the key-lever. A comparison between the clavichord and fortepiano is instructive: the distance of travel of the tangent to the string is about a quarter of the travel of the fortepiano's hammer head to the string. Furthermore, and of more significance, there is a multiplication of the acceleration in the fortepiano (due to the escapement) of five to six times. In the clavichord, this multiplication is at most about twice. Thus, the fortepiano is much louder than the clavichord.

In determining the vibrating length of the strings, the tangent (while in contact with them) also determines the resulting pitch in much the same way as a guitarist determines the pitch that a string will sound by pressing it against one or another of the frets on the fingerboard of his instrument. Accordingly, by positioning a series of clavichord tangents so that they will strike the same pair of strings at different points along their length, a series of different notes can be sounded. This possibility was exploited in all the earliest clavichords, which are termed 'fretted' to distinguish them from 'unfretted' instruments (of the late 17th century and since), in which each note is produced by its own pair of strings. The use of only a single pair of strings to serve several keys has the obvious advantage of reducing the number of strings on the instrument, which permits in turn a lighter and simpler case to withstand their tension. Moreover, the smaller number of strings exerts a smaller downward pressure on the soundboard, which can then vibrate more freely. Finally, the smaller number of strings permits the instrument to be tuned more rapidly and more easily (see §2 below).

A disadvantage of the fretted clavichord comes from the fact that a single pair of strings can sound only one note at a time, making it impossible to play chords involving two notes whose tangents strike the same pair of strings (only the upper note will be heard, usually with an unpleasant clicking from the tangent of the lower note). The same factor makes it necessary to avoid a complete legato in playing descending scale passages on a fretted clavichord when the scale includes consecutive notes produced from the same pair of strings.

The clavichord is not easy to play well: since the pressure exerted on the keys affects the pitch of the notes while they are sounding, the performer must maintain an even pressure after striking and must think about the notes all the time that they are sounding and not merely at their beginning and end.

2. 15TH AND 16TH CENTURIES

The earliest known appearance of the term 'clavichord' occurs in Eberhart von Cersne's *Minne regel* of 1404 (ed. F. X. Wöber, 1861), and the oldest known representation of the instrument is in a carved altarpiece dated 1425 from Minden in north-west Germany.

As described, all the earliest clavichords were so designed that each pair of strings was struck by the tangents of several keys; hence the keys had to be curved or bent laterally (see fig.39, p.144) so that their tangents would touch the strings at appropriate points along their length. These points were determined by the monochord measurements for the intervals, and clavichords were, accordingly, called *monochordia* by many 15th- and 16th-century writers. It was, of course, necessary to have more than a single string or pair of strings, as it would otherwise have been impossible to sound more than one note at a time; nevertheless in the earliest clavichords the strings were all of the same length and all tuned in unison, so that these instruments were, in effect, a series of identical monochords built into a single box, and the positions of the tangents along the strings were determined as if there had been only a single string. This may be seen clearly in the directions for laying out clavichords that occur in a number of 15th-century sources, the first step in which is to divide the total string from the bridge to the tangent of the lowest key into several parts corresponding to notes covering virtually the entire compass of the instrument.

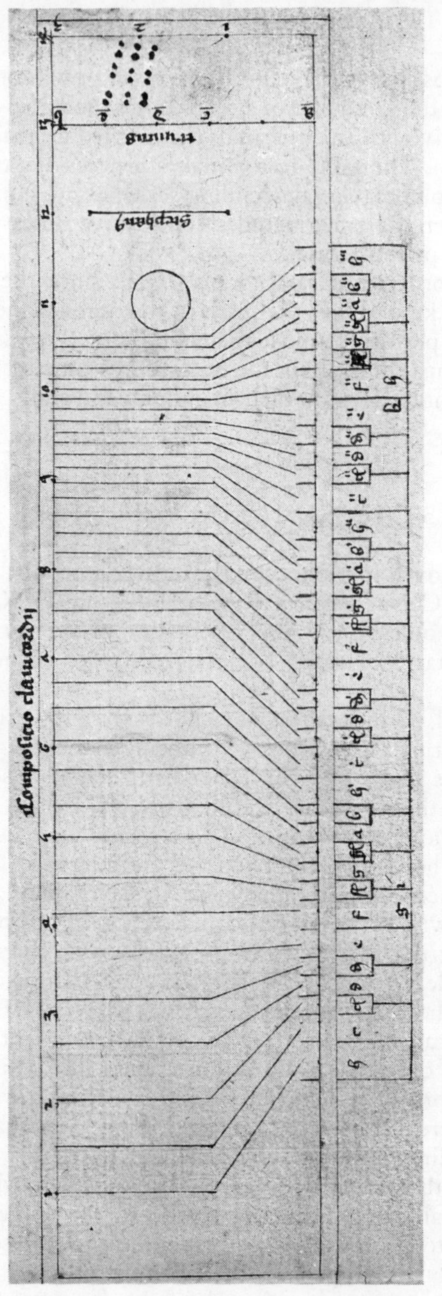

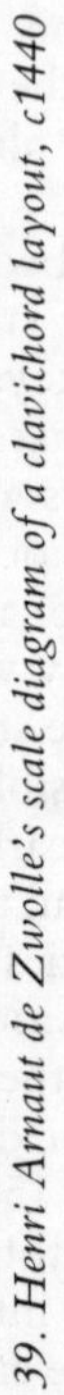

39. *Henri Arnaut de Zwolle's scale diagram of a clavichord layout, c1440*

Instructions of this kind accompany the scale diagram of a clavichord layout (fig.39) given in the manuscript treatise (*c*1440) by Henri Arnaut de Zwolle, which is the earliest source of information on the way in which the tangents were apportioned to the strings of the instrument. The tangents for the first seven pairs of strings on Arnaut's three-octave instrument (after this point there seems to be an error in the manuscript) are assigned in fours and threes corresponding to the following groups of notes: *B–d*, *e♭–f♯*, *g–a*, *b♭–c♯'*, *d'–e'*, *f'–g♯'* and *a'–b'*. Except in the first of these groups, four keys are served by the same string only when the outermost notes form an augmented 2nd, and whenever a fourth key would produce the interval of a minor 3rd (such as *d'–f'* or *a'–c''*) the number of tangents allotted to a single pair of strings is reduced to three. The result of this arrangement is that virtually any consonant chord can be played, as its constituent notes will always be sounded from different strings, and the only notes that cannot be sounded simultaneously are those forming the dissonances of a minor, major or augmented 2nd. The clear implication of having the tangents allotted in this pattern, which continued in use on some clavichords until the end of the 17th century, is that even as early as the mid-15th century, keyboard players expected to be able to play consonant chords with complete freedom.

In addition to having all its strings of equal length and tuned in unison, the mid-15th-century clavichord differed from those known today in having its soundboard located near the bottom of the instrument and extending underneath the keys. The bridge over which the strings passed was, accordingly, quite high, and 15th-century representations are unanimous in indicating that it had a shape resembling that of the bridge of a viol.

Despite the ease of laying out a clavichord of the kind just described, and the ease in tuning suggested by the intriguing possibility of simply removing the cloth from between the strings, plucking the strings, and bringing them into unison, instruments of this type had at least one important disadvantage. Because on such an instrument the sounding length of the strings had to double for each octave, the sounding length of the string for the lowest note of an instrument with a three-octave compass like Arnaut's had to be eight times as long as that for the highest note. This meant that some keys had to be sharply bent, and any appreciable increase in range would have required keys too bent (in the tenor) or too thin (in the treble) to function.

40. *Intarsia (1479–82) of a clavichord in the ducal palace at Urbino*

It was not really necessary, however, to leave the same amount of space between keys playing on different strings that would be required if they played on the same strings. For each change of string, a space could be eliminated by placing the highest key playing one pair of strings immediately next to the lowest key playing the adjacent pair of strings. This, however, sacrifices the unison tuning of all the strings because the pair of strings of the lower group of keys would be tuned somewhat lower to compensate for this group being relatively closer to the bridge. Carrying this principle to its logical conclusion, the inordinately wide spaces between each of the keys in the extreme bass could all be avoided by giving each key its own pair of strings. Thus, abandoning unison tuning of all the strings would make it possible to produce a far more compact three-octave instrument with less sharply cranked keys, or to make a workable instrument of wide compass.

The earliest representation of the newer type of instrument is the intarsia of a clavichord in the ducal palace at Urbino (fig.40), made between 1479 and 1482, which shows an instrument with a four-octave range *F* to *f'''* (but without *F*♯ and *G*♯), sounded from 17 pairs of strings. The first five notes (*F*, *G*, *A*, *B*♭, *B*) each have strings of their own, and the remaining three and a half octaves are accommodated on 12 pairs of strings, which, with the exception of the highest pair, have their tangents arranged in threes and fours in the manner described for Arnaut's instrument (i.e. no group of four tangents encompassing a consonant interval). The Urbino example avoids Arnaut's anomalous inclusion of his bottom note (*B*) in a group of four tangents (*B* to *d*) but includes the top note (*f'''*) in a group of four rather than giving it an extra pair of strings to itself.

Clavichords of this kind were probably still being made in the 1530s, when an interesting representation attributed to the Flemish artist Jan van Hemessen was painted (fig.41, p.149). This instrument still has a low soundboard, and its range of *F* to *a''* (without *F*♯, *G*♯ and *g*♯) is exactly that required by Hugh Aston's famous *Hornpype*, Attaingnant's *Quatorze gaillards* (1531) and Gardane's *Intavolatura nova di varie sorte de balli* (1551).

By the date of the earliest surviving clavichord, a hexagonal instrument made by Dominicus Pisaurensis (Domenico da Pesaro) in 1543 (fig.42, p.150), a second significant change in clavichord design had occurred. The soundboard of Dominicus's instrument does not run beneath the keys, but, as in all other surviving instruments, it is at a higher level than the keys and placed entirely to their right. As a result there is no longer the need for the large viol-type bridge of the earlier clavichords. This change would undoubtedly have benefited the volume of the treble notes, since a large mass (i.e. a bridge) on the soundboard reduces the sound output. It was no doubt also a technical solution to the problem of arranging that the lower keys played adequately long strings, since short strings have a poor sound in the tenor and bass. At the same time it enabled the sideways splay on the keys to be reduced, since instead of splaying the key to the left separate bridges were added for the tenor and bass, which were further from the keyboard than the treble bridge. The solution adopted by Dominicus Pisaurensis and a number of other 16th-century makers, still to be seen in the clavichord 'italienischer Mensur' depicted by Praetorius (see fig.43, p.153) and that shown by Mersenne (1636–7), was to divide the bridge into segments. One segment, carrying the 11 pairs of strings that serve more than a single key, is set near the left-hand edge of the soundboard; two others carrying five and six pairs respectively (the 11 pairs of strings serving the first 11 notes in the bass) are set farther to the right. The soundboard slopes downwards to the right, so that the strings leave the low treble bridge at a fairly sharp angle and exert a downward pressure on the bridge that ensures their not being lifted from the bridge when they are struck by the tangents. Some soundboards were flat over their whole length and had bars over the strings (between the bridge and tuning pins) in order to ensure adequate pressure on the bridge. This arrangement has the disadvantage that the length of string between bridge and tuning pin cannot vibrate in sympathy with the struck strings. Thus, the overall sound of such an instrument is somewhat duller.

Like the case of an Italian virginal, the case of an Italian clavichord may be made of thin wood strengthened by elegant mouldings, or of thicker softwood with a cypress lining and half-moulding to counterfeit the appearance of a thin-cased instrument in a protective outer case. The similarities in style between German and Italian instruments make it unclear whether all the surviving examples of clavichords with segmented bridges and relatively thin cases are actually of Italian origin, and it is possible that two well-known rectangular examples in the Musikinstrumenten-Museum of the Karl-Marx-Universität, Leipzig, are in fact German.

41. 'Girl Playing a Clavichord': painting (1534) attributed to Jan van Hemessen

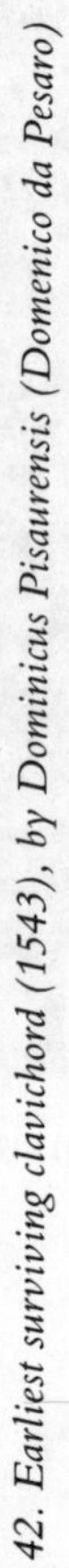

42. Earliest surviving clavichord (1543), by Dominicus Pisaurensis (Domenico da Pesaro)

The sound of the surviving 16th-century clavichords is surprisingly loud and virginal-like; they are sensitive and exciting instruments to play, ideally suited to pieces like the dances found in the Attaingnant and Gardane collections (both of whose title-pages mention the clavichord) as well as the elaborate intabulations of vocal works that seem to have comprised most of the balance of the 16th-century keyboard player's repertory.

Despite the 16th-century title-page references to the clavichord, implying an importance comparable with the harpsichord and virginal, the clavichord at this period seems to have been thought of primarily as a teaching and practice instrument. Many early references cite it in these connections, extolling its virtues in developing facility and a proper touch which might then be transferred to other instruments, notably the organ. Its advantages as a practice instrument were outstanding, especially its cheapness; it gave organists the opportunity of practising at home instead of in an unheated church in winter, and eliminated the need for an assistant to pump the bellows. A number of writers even praise the softness of the clavichord's sound as being of advantage when practising.

A further advantage of the clavichord was that it tended to stay in tune and was easy to put back into tune when necessary. With most of the strings serving several keys, there were fewer strings to tune, and as the intervals between notes sounded from the same pair of strings were 'built in' by the spacing of the tangents, much of the difficulty in setting the base tuning of the instrument was eliminated and a means of checking the accuracy of one's attempts was more quickly arrived at. One highly ingenious system set down by Correa de Arauxo (1626) makes use of the alternating fretting pattern of threes and fours (which produces different groupings of notes in adjacent octaves) to tune the instrument entirely by alternating upward and downward octaves, without having to use 5ths or 4ths. Although clavichords are usually strung with brass wire throughout, it seems most likely that many of the 16th-century Italian clavichords were intended to be strung with iron wire.

Furthermore, the scales of these instruments were quite short: the clavichord of Dominicus Pisaurensis has a *c''* of 23.2 cm. The shortness of scale arises to some extent from practical necessity, since with three- and four-note fretting down to *b♭* a long scale would produce unacceptably large spacing between the keys; this is mechanically unsatisfactory for the guide system. Considering the usual constructional practice for 16th-century Italian virginals and harpsichords, it seems probable that the Dominicus clavichord was intended for iron stringing in the treble with brass lower down. This gives an advantage to the bass strings, which stand at a higher pitch than if the instrument had been strung in brass wire. Furthermore, it would seem that this scale was chosen so that the instrument would be a 5th above the commonly used Venetian scale of *c''* = 35 cm (see Chapter One, §2(i)). What strengthens these probabilities to a near certainty is the report of Virdung (1511) that some of the 'newer' clavichords were strung with iron in the treble and brass lower down.

Several of the 16th-century instrument treatises and methods provide considerable information about the clavichord. Virdung mentioned clavichords with all their strings tuned in unison and showed a clavichord keyboard with a range of just over three octaves, *F* to *g''* without *F*♯. He stated that the 'newer' clavichords – which might have a range as great as four octaves – might be triple-strung to avoid problems if a string broke during playing. The use of strings made of different types of wire makes it clear that these larger instruments did not have all their strings tuned in unison. Virdung also wrote that the four-octave instruments had pedals hanging from their lowest keys and that such clavichords had extra strings which were not struck by any tangents. No surviving clavichord has sympathetic strings, but one can imagine how they might enhance the tone with a halo of sustained sound.

Both Bermudo (1555) and Santa María (1565) included diagrams of a clavichord keyboard extending to *a''* in the treble (including *g*♯*''*) and *C* in the bass by means of a short octave, like that in the 1543 clavichord by Dominicus Pisaurensis. Santa María provided a highly detailed discussion of clavichord technique, but neither he nor any of the other early writers alluded to the instrument's expressive possibilities or suggested that the clavichord had any special musical potential of its own.

3. 17TH CENTURY

Inventories, account-book entries and other writings suggest that clavichords were common all over Europe in the 16th

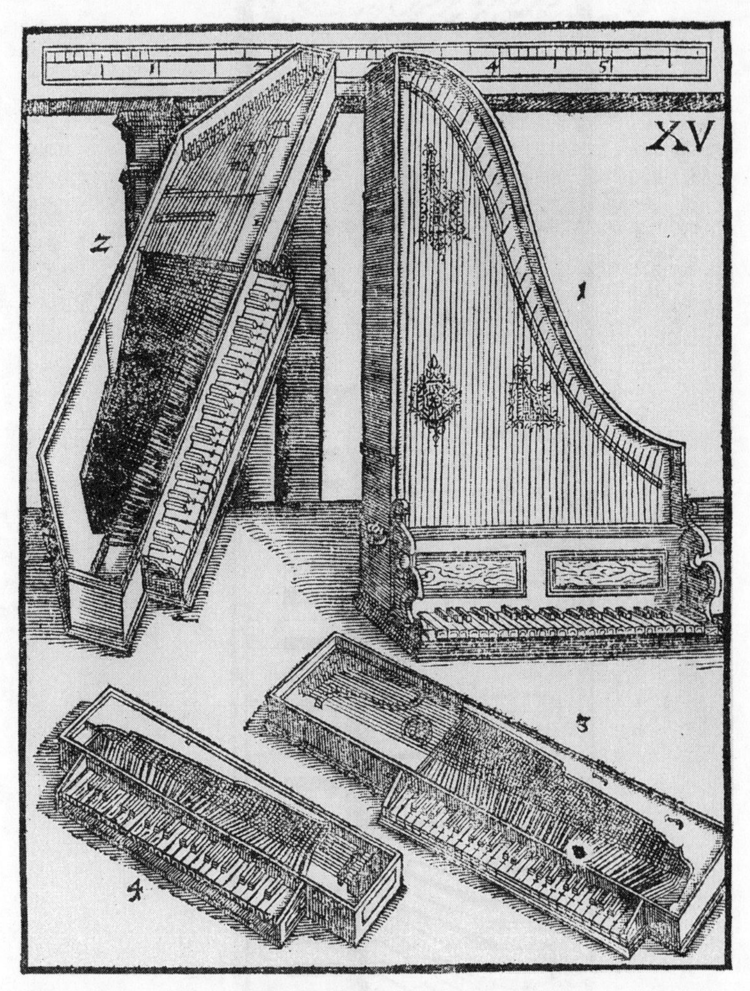

43. Polygonal Italian clavichord (2), with two rectangular clavichords, 'Gemein' (4) and 'Octav' (3), and a clavicytherium (1): woodcut from Praetorius's 'Syntagma musicum' (2/1619, plate XV)

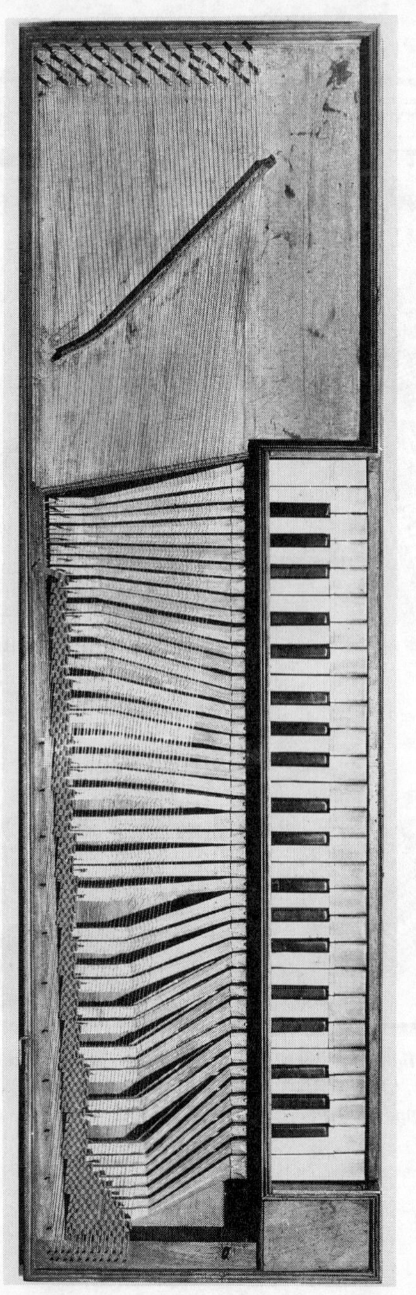

44. ?German fretted clavichord, c1700

century but that in time the instrument became appreciably less popular outside Germany, Spain, Portugal and Scandinavia. As early as 1547 the collection of instruments owned by Henry VIII included only two clavichords in contrast to 30 'virginalles' of various types, 24 'regalles', 'portatives' and 'organes' and three 'virgynalles' and 'regalles' combined. In France, Mersenne (1636–7) provided a description of a clavichord so vague and inconsistent that one wonders if he had ever actually seen one; and the instrument in his illustration, despite its vaulted lid and alleged chromatic bass octave, looks more like the hexagonal, thin-cased Dominicus Pisaurensis instrument of 1543 set into a protective outer case than it does an instrument made either in France or in the 17th century. In the Low Countries the clavichord appears in a small number of 17th-century paintings. Relatively unsophisticated types were discussed by both Douwes (1699) and Blankenburg (1739), the latter specifically referring to the clavichord as 'the organist's study instrument'; but, although Douwes described a pedal clavichord (an instrument equipped with a pedal-board like that of an organ) and Blankenburg mentioned a two-manual instrument, neither devoted as much space to clavichords as to quilled instruments, and Verschuere-Reynvaan, when he copied Douwes's already archaic text (1789; 2/1795), added nothing but an illustration of a pedal clavichord. This suggests that the clavichord was well out of the mainstream in the Low Countries, which is hardly surprising in view of the great importance of Flemish harpsichord building.

The surviving instruments support the written evidence: most of the surviving clavichords of the 17th, 18th and 19th centuries are German, the rest being mostly Scandinavian and Iberian. Only a few clavichords that may have been made in the Low Countries are known; two examples may possibly be English but no French clavichord is extant. Accordingly, the history of the clavichord from the 17th century onwards is largely the history of the clavichord in Germany.

In 1618 Praetorius gave a good idea of the clavichord's importance in Germany at the time, predictably citing its value as a practice instrument, and one of the woodcut plates (see fig.43) later issued to illustrate his text shows three representative types: a polygonal Italian instrument which Praetorius said had been brought to Germany 30 years before, and two smaller rectangular instruments presumably of German make. These two have four-octave keyboards *C/E* to *c'''*, and the smaller instrument, designed to be tuned an octave above normal pitch, appears to have a segmented bridge and a low soundboard; the larger one, labelled 'ordinary clavichord', is like many surviving instruments except that its keyboard projects rather than being inset as became standard later in the 17th century. (It is shown reversed in the engraving.) The greatest change to be seen in this instrument is that instead of a segmented bridge it has an S-shaped bridge in one piece, its treble end close to the left edge of the soundboard and its bass end towards the opposite edge.

Each of Praetorius's clavichords is equipped with a tangent rail, a flat triangular board that might rest on the damped section of the strings. It is possible that such boards obviated the need for listing woven between the strings, which would have simplified the replacement of broken strings, but experience with the tangent rails on surviving original instruments and on 20th-century examples suggests that listing is still required and that the principal function of the tangent rail is to bear down on the strings close to the point at which they are struck by the tangents. This has the effect of making it harder for the tangents to lift the strings, so that the notes speak more clearly and one can play the instrument more loudly before the pitch is unduly raised. Unfortunately, the feel of the keys while playing may lose its normal yielding quality and it may thus become more difficult to control the continuing tone by variations in finger pressure, with the result that the sound of the instrument may seem as hard and unyielding as its touch. It is difficult, though, without knowing the exact string sizes used by the maker, to make a final judgment on the touch and musicality, since this is strongly dependent on the stringing: if the string tension is too little, the pitch will sharpen too much when playing; if too much tension is employed the keys will seem to bounce back uncontrollably after the tangents strike the strings. The clavichord maker has therefore to balance a number of factors in producing a musical instrument.

Surviving 17th-century German instruments show a number of variable features. In many cases, the bridge is not S-shaped like that shown by Praetorius but rather is straight or only gently curved, and placed obliquely on the soundboard (see fig.44); other instruments have a bridge with a single sharp curve at the treble end, a shape seen commonly on 18th-century instruments as well. The system of fretting in fours and threes began to be replaced later in the 17th century by systems involving only threes and eventually by another system of great importance involving the allotment of no more than two keys to any pair of strings (see below). The increased number of strings required a rearrangement of the tuning pins, from their position at the right-hand end of the case to a slanting line from the front right-hand corner, or, more usually, to a short perpendicular section at the front right-hand corner of the case merging with a longer oblique section running across the back right-hand corner.

The single greatest advance in clavichord design during the period immediately following Praetorius was the adoption of a layout that included bringing the front of the case forward so that the keyboard would be inset rather than projecting. This resulted in an enlargement of the soundboard area and made it possible to use diagonal stringing, as the front part of the bridge carrying the bass strings could now be brought forward on to the soundboard. As a result of diagonal stringing, the row of tangents could follow a line far more nearly parallel to the front of the keys, and the keys could be made somewhat shorter and more uniform in length, thereby improving the touch of the instrument. Diagonal stringing also brings with it a disadvantage: the string tension twists the case. Some old instruments show considerable case distortion.

An instrument of this kind with a *C/E* to *c'''* four-octave compass can be quite small – a typical example dated 1652 is only 109 × 32 × 10 cm and uses only 20 pairs of strings. The first seven notes in the bass each have their own pair of strings, there are two groups of two keys each, and the rest of the compass is fretted in threes. These 17th-century clavichords retain much of the bright assertive tone of 16th-century examples, while their improved touch tends to make their sound rather more flexible.

By the 1690s instruments of this size or even a trifle smaller were being made with no pair of strings struck by more than two tangents. The first five to eight keys would each have their own pair of strings, and in a version of the system used in Germany the remaining notes would be disposed as follows: all C's paired with C♯, all D's unpaired, all E's paired with E♭, all F's paired with F♯, all G's paired with G♯, all A's unpaired, and all B's paired with B♭. The advantages of this arrangement to the performer are enormous. There are still relatively few strings to go out of tune – a maximum of 30 pairs for a four-octave instrument – and to some extent the simplified tuning of earlier instruments remains; however, the performer gains virtually all the freedom in playing dissonant chords and legato scales that he would have if each key had its own pair of strings – as long as he remains within the bounds of those tonalities employing no more than two flats or three sharps, since he will not then find himself using any of the paired notes simultaneously, and rarely in immediate succession.

At this time, however, clavichords were already being made in which each key was provided with its own pair of strings, and Johannes Speth, in the preface to his collection of keyboard pieces, *Ars magna consoni et dissoni* (1693), specified that the music should be played on a virginal or a clavichord 'so made that each key has its own strings and not so that two, three and up to four keys touch a single string'. (Yet there is nothing apparent in Speth's music itself to require such an instrument rather than one fretted in pairs.)

The structure of a 17th-century clavichord tends to be very simple. The sides of the case, usually dovetailed together at the corners, are mounted on a solid bottom, which is not stiffened by any transverse member other than the balance rail of the keyboard. The wrest plank is attached to the right-hand end of the case and in some examples does not run all the way down to the bottom; the hitch-pins for the left-hand ends of the strings are driven into hardwood liners along the back and left-hand ends of the case, and the slotted diapason and a padded rail on which the keys rest are placed immediately in front of the back liner. A small toolbox is usually provided in the space to the left of the keyboard, and a major brace supporting the left-hand edge of the soundboard runs from the front of the case to the back at the right of the keyboard. This brace is often pierced by a soundhole, or there may be a soundhole decorated with a rose in the soundboard itself.

4. 18TH-CENTURY GERMANY

Occasional clavichords from the last years of the 17th century have divided sharps in the lowest octave to provide the *F*♯ and *G*♯ omitted in the normal *C*/*E* short octave, and some instruments were made with chromatic bass octave (still occasionally omitting the *C*♯) around 1700. A further downward extension to *G*′, corresponding to that seen in harpsichords in Flanders and France by the mid-17th century and in Germany by 1710 at the latest, did not occur as early in clavichords, which retained the C-based keyboard of the organ until the mid-18th century. An upward extension of the range seems to have been similarly delayed, although an isolated instrument built by Baumgartner in Bolzano and dated 1683 has an atypical chromatic compass of *C* to *f*′′′.

The complete emancipation of the clavichord's keyboard range from that of the organ seems to have occurred in the 1740s, well after appreciation of the unique virtues of the clavichord as contrasted with those of the harpsichord and organ had begun to appear in writings on music. As early as 1713 Mattheson singled out the clavichord as 'beloved above all' other keyboard instruments and declared it superior for performing 'overtures, sonatas, toccatas, suites, etc' because it permits one to produce a 'singing style' of performance. The importance of such a style is emphasized in the title of Bach's Two- and Three-part Inventions (1723), where the pedagogical purpose of the music is specified as being 'above all to achieve a cantabile manner of playing'; it is in just this respect that the clavichord excels in contrast (to quote Mattheson again) to the 'always equally loud, resonant harpsichord'. This emphasis on singing style, with dynamic nuance explicitly demanded (and some pitch flexibility probably implied as well), heralds the period in which the clavichord began to have a literature of its own and in which large clavichords were first made.

The earliest surviving instruments of this type were all made in Hamburg. Two examples built in the 1720s by Johann Christoph Fleischer and Hieronymus Albrecht Hass are approximately 150 cm long and nearly 45 cm from front to back. The range of these instruments is *C* to *d*′′′ and both are fretted in pairs. The next surviving Hass clavichord (made in 1740) is similar, but in 1742 he built the earliest surviving five-octave instrument, an unfretted example with a range (*F*′ to *f*′′′) equal to that of the largest harpsichords of the period. Such clavichords became increasingly common in the 1750s, although Hass continued to make an occasional instrument of shorter range and his son even built a fretted example as late as 1768. They have, in addition to the usual unison pair of strings, a third

45. *Unfretted German clavichord by J. A. Hass, 1763*

string tuned an octave higher in the bass and extending to *c*. These strings have a short, straight bridge of their own and are held by hitch-pins set in the soundboard, rather like the 4′ strings of a harpsichord; their tuning pins are at the left-hand end of the case. The effect of such strings, which seem to be a particular feature of the Hamburg instruments and those of makers strongly under the Hamburg influence, is to brighten and lend definition to the bass, but there tends to be an audible break at the point at which they are no longer present.

The Hass clavichords are beautifully made and often exquisitely decorated with chinoiserie (see fig.45,) but, like the Hass harpsichords, they can be a trifle disappointing musically. The very long key-levers, necessitated by the wide total string-band (i.e. the distance from the lowest to the highest string) and correspondingly broad case, give them a certain feeling of sluggishness instead of the firm, positive feel of earlier clavichords; this adds to the difficulty of playing them well. Their tone is somewhat veiled and less assertive although undeniably beautiful.

Throughout the 18th century small, unpretentious instruments, many of them fretted, continued to be made in Germany in addition to the large unfretted ones (see fig.46, p.162). Some of these were designed as travelling instruments, like the one made by Johann Andreas Stein for the Mozart family in 1763 and the small ones belonging to Beethoven and Grétry. But certain makers, most notably C. G. Hubert of Ansbach, specialized in making intermediate-sized fretted instruments of the highest quality, with a four-and-a-half-octave range of *C* to *f‴* or *C* to *g‴*. Although these instruments might be about as long as an unfretted instrument of comparable range, permitting bass strings of adequate length for good sound, the elimination of the notes below *C* and the use of fretting by pairs yields a very narrow stringband and permits the keys to be short, which provides a snappy action and superb, sensitive touch. In addition, the smaller number of strings reduces the load on the soundboard, yielding a clearer, more singing tone.

With all these advantages, one might well wonder why unfretted clavichords eventually superseded fretted ones. In fact, even on a clavichord fretted only in pairs a performer encountered difficulties if he strayed beyond the bounds of the tonalities having no more than two flats or three sharps, for he might then have to play two notes sounded from the same pair of strings

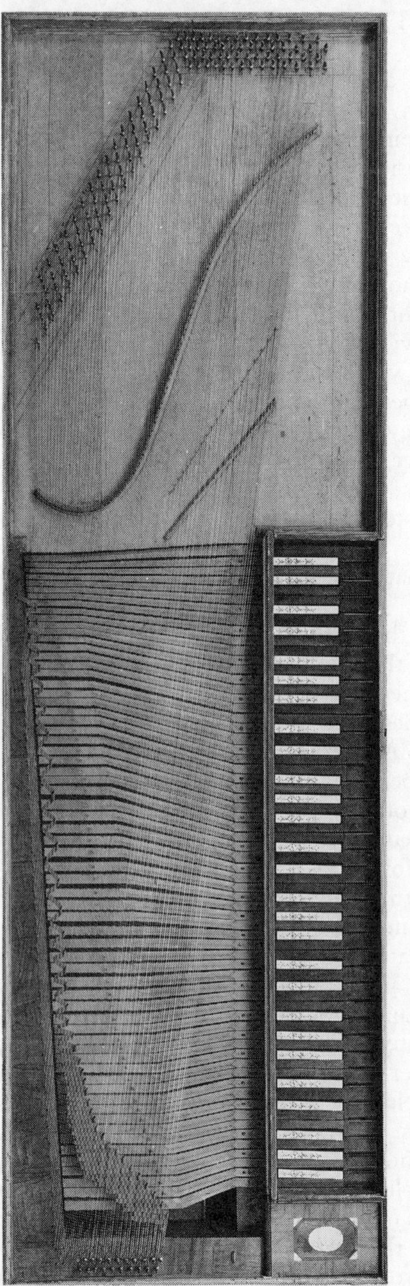

46. Unfretted clavichord by Barthold Fritz, Brunswick, 1751

either simultaneously or in immediate succession. Consequently, his freedom in playing chords and producing a legato touch in all keys was circumscribed. Moreover, the fact that the size of the intervals between notes sounded from the same pair of strings could be altered only by bending the tangents inwards or outwards (and then only within fairly narrow limits) made it impossible for the performer to alter the tuning at will to favour different tonalities. This was a particularly serious limitation for the musician using mean-tone tuning, as this system set very narrow restrictions on the number of tonalities that could be played in acceptable intonation without retuning one or more of the sharps or flats in each octave. Thus, it seems that the clavichord fretted in pairs fell into disuse, despite its many virtues, because musicians wished to have the freedom in tuning and in playing in all keys that could be achieved only on an instrument having each key served by its own pair of strings.

Jacob Adlung (1758 and 1768) gave the most detailed surviving account of the clavichord in Germany in the third quarter of the 18th century, including such types as the pedal clavichord and the cembal d'amour (see below), and described a variety of devices applied to the instrument to give it a range of different tone-colours or otherwise enlarge its capabilities. Several of these involved enabling the keyboard to be moved inwards or outwards: the keyboard could be moved a small distance so as to cause the tangents to strike only one of the two strings in each pair; or it could be pulled far enough forward to have the tangents strike the strings of a different pair rather than the normal one so as to achieve automatic transposition. Another was the use of wide tangents with one half of the striking surface covered with leather to provide the muting effect of the buff stop on a harpsichord (an effect achieved otherwise on one or two surviving clavichords by interposing a dentated strip of leather between the tangents and the strings).

The cembal d'amour was invented by Gottfried Silbermann in Freiberg, Saxony, in the first quarter of the 18th century. Although the name 'cembal d'amour' might appear to imply that the instrument was a harpsichord with sympathetic strings (by analogy with the viola d'amore) it was in fact not a harpsichord at all. Rather, it was a clavichord with strings of approximately twice the normal length that were struck by the tangents precisely at their midpoint; the name seems to derive from the fact that the instrument's sound blended well with that

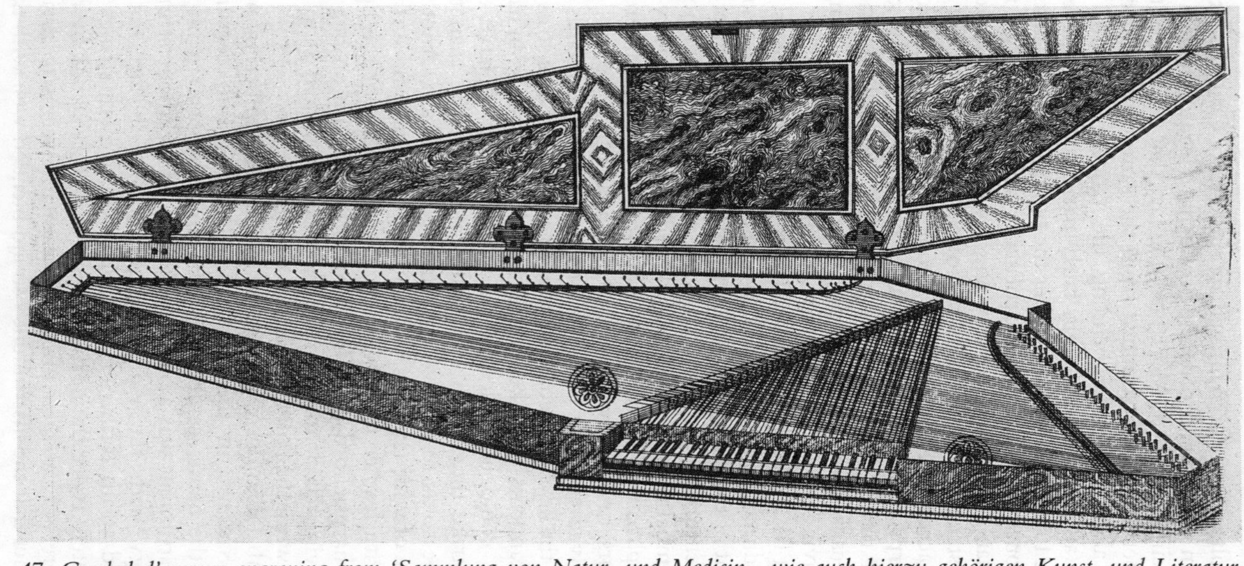

47. Cembal d'amour: engraving from 'Sammlung von Natur- und Medicin-, wie auch hierzu gehörigen Kunst- und Literatur-Geschichten' (June 1723)

of the viola d'amore. The invention of the cembal d'amour is announced in the *Sammlung von Natur- und Medicin-, wie auch hierzu gehörigen Kunst- und Literatur-Geschichten* (Breslau and Leipzig) for July 1721. Although no example of the instrument has survived, it is depicted in the June 1723 issue of the *Sammlung* (see fig.47,) and in a coloured drawing among the papers of Johann Mattheson; in J. F. Agricola's annotations to Adlung's *Musica mechanica organoedi* (1768) there is a description and diagram of it. J. N. Forkel in 1781 stated that the tone of the cembal d'amour sustained longer than that of an ordinary clavichord, was far louder (though not so loud as a harpsichord, but, rather, 'midway between the two') and that its dynamic range was also greater than that of an ordinary clavichord though far inferior to that of a fortepiano.

The instrument had an irregular form dictated by the presence of two bridges, each resting on its own soundboard, one behind and to the left of the keyboard and the other in the normal position to the right of the keyboard. The two segments of the strings vibrated independently when the tangent struck and, since the segments were of equal length, they sounded in unison and produced a louder tone than that of an ordinary clavichord, in which only the right-hand segment is allowed to sound, while the left-hand segment is damped by strips of cloth woven between the strings. Since this cloth, however, also serves to damp the strings as a whole when the key of an ordinary clavichord is released, Silbermann had to compensate for its absence in the cembal d'amour by devising another means of damping. The information given in *Musica mechanica organoedi* suggests that each of the strings of the cembal d'amour rested on the forward-projecting prongs of a U-shaped block covered with cloth. The tangent rose between the prongs of this U, lifting the strings from the cloth and undamping them at the moment that they were struck by the tangent.

Because of the double length of its strings the cembal d'amour tended to be quite large, even when its range did not extend below *C*, as in the *Sammlung* and Mattheson's drawing. This disadvantage was partly overcome in an interesting variation on the instrument devised by the mathematician Leonhard Euler and described in the supplement to the third edition of the *Encyclopaedia Britannica* (Edinburgh, 1800) in which the string segment at the left was, for most of the range, only half the length of that at the right and sounded an octave above it, like the 4′ strings in the bass of some large clavichords or the 4′ register of a harpsichord. The author of the *Britannica* article, John Robison, praised the sound of Euler's instrument, but claimed that in the bass (where, in the interest of saving space, the length of the left-hand segment was made a small fraction of the right-hand segment and where the strings were overspun rather than plain) the tone was inferior and that 'the instrument was like the junction of a very fine one and a very bad one and made but hobbling music'. Several other builders are known to have made cembals d'amour. Johann Ernst Hähnel of Meissen equipped his instruments with a device for silencing the strings on either side of the tangents if desired, and also a so-called pantalon stop (see below).

Adlung mentioned the possibility of placing a special bridge between the tuning pins and the regular bridge at a distance from the latter corresponding to half the sounding length of the strings, so as to leave to the right of the bridge an undamped portion of string which would resonate automatically at the octave above the note played. He referred to an apparatus known as the 'pantalon' stop. This stop (described by Adlung in 1758) was named after the pantaleon or pantalon, a large dulcimer invented by Pantaleon Hebenstreit (1667–1750), the tone of which was enhanced by the resonance of undamped strings and which was capable of producing dynamic nuance. The stop consists of a series of tangent-like brass blades set in a movable bar so that all of them can be raised at once by the action of a stop-knob. When raised, these blades touch the strings immediately to the right of the point at which they are struck by the tangents carried by the keys. When the keys are released the strings rest on the pantalon-stop blades, which continue to separate the sounding part of the strings from the cloth damping woven between them at their left-hand end: hence the strings continue to sound instead of having their vibrations damped out as soon as the key is released. In addition, the strings vibrate sympathetically with other notes being played, so that the effect produced by the stop is essentially that of a hand-operated damper-lifting mechanism like that on many early square pianos. In the few surviving clavichords equipped with a pantalon stop, it is divided so that it may be used in only the treble or bass.

Adlung stated that the pantalon stop made the sound of a clavichord 'cutting and bright' and even enabled a player to use the instrument to perform chamber music with recorders or an unmuted violin. (He went on to suggest that if the undamped strings ring on too long the pantalon-stop blades may be topped with leather or cloth to make the sound die away more quickly.) Surviving examples of the pantalon stop do not seem to justify completely Adlung's enthusiasm for the device. Their most annoying characteristic is that, since the pantalon-stop blades cannot touch the strings at exactly the same point as the tangents do, there is a detectable rise in the pitch of a note after the key is released.

One contraption found on surviving clavichords brings the backs of the keys nearer to the strings, thereby shortening the key-stroke and reducing the loudness of the tone. Like the muting effect produced by interposing a strip of leather between the tangents and the strings, this device may well be an example of the influence of the technology of the square piano on the older clavichord.

If Adlung – who was an amateur clavichord builder himself – gives the impression that the 18th century's love of gadgetry was about to transform the clavichord into something alien to its essential nature, C. P. E. Bach writing at the same period (*Versuch*, i, 1753) made it clear that other musicians were not losing sight of the true character of the instrument. He mentioned no stops or special effects, speaking rather of the instrument's ability to render all shades of dynamic nuance and to produce a vibrato (*Bebung*) and portamento, and concentrating on such essentials of a good clavichord as an even, responsive touch, a 'sustaining, caressing tone', and a range of at least *C* to *e'''* (the highest notes being required when playing music intended for other instruments – presumably the violin or the flute). C. P. E. Bach 'could not bear' octave strings in the bass of a clavichord and made no mention of unfretted instruments, on which Johann Samuel Petri (2/1782) and Daniel Gottlob Türk (1789; ed. Jacobi, 1962) insisted. Bach's personal instruments by Silbermann and Friederici were certainly unfretted none the less.

C. P. E. Bach's views concerning the clavichord are especially significant, as in addition to writing the most influential treatise on 18th-century keyboard playing he was certainly the most important composer to conceive his music in terms of the clavichord. The appearance of *pianissimo* indications as well as *forte* and *piano* in certain sonatas published in the 1740s, despite title-pages describing them as 'per cembalo', suggests that he may have had the clavichord in mind for these works. Explicit *Bebung* indications appear in one of the compositions written to illustrate his *Versuch* and in the first of the *Kenner und Liebhaber* collections (1779). (See also Chapter Five, §4.)

Later German writers were utterly unrestrained in their praise of the clavichord, especially as a vehicle for the most intense and private personal expression. Schubart wrote (in 1786) that a clavichord 'made by Stein, Fritz, Silbermann, or Späth is tender and responsive to your soul's every inspiration, and it is here that you will find your heart's soundboard . . . Sweet melancholy, languishing love, parting grief, the soul's communing with God, uneasy forebodings, glimpses of Paradise through suddenly rent clouds, sweetly purling tears . . . [are to be found] in the contact with those wonderful strings and caressing keys'. The heightened sensibility and sentimentality evident in Schubart's rhapsodizing are closely attuned to the *Sturm und Drang* and *Empfindsamkeit* styles of the second half of the 18th century, part of the special climate in which the clavichord had its great flowering of popularity and in which all of its special literature was created.

During the last quarter of the century some clavichord makers seem to have been working towards an instrument with the intimate sensitivity and flexibility so highly praised by Schubart, while others seem to have been working towards a louder sound more appropriate for the piano repertory that was developing at this time. Virtually all the important German clavichord makers were by this time building pianos as well. F. C. W. Lemme, who developed a clavichord with a case having rounded ends for which he claimed 'an uncommonly beautiful tone', advertised in 1802 that he made 'large grand pianos in the style of the English masters . . . another large grand piano . . . square pianos . . . and clavichords of all kinds, all of them unfretted'. (He listed no fewer than 14 different models.)

The tendency to build piano-like clavichords was increasingly reflected in the sound and massiveness of structure of many instruments made in the 19th century by such German builders as Voit and Schmahl and the Scandinavian makers, some of whom were still producing clavichords in the 1820s. The structure of a large 18th-century clavichord differs from that of a 17th-century instrument in a number of details. A certain massiveness was required to withstand the tension imposed by the greater number of strings used for the expanded range of the later instruments and because they were generally unfretted. In the 1790s a diagonal brace running parallel to the strings was often attached to the inside of the bottom to stiffen it and thus help prevent twisting of the case. The diagonal part of the wrest plank was not usually thick enough to reach the bottom of the case but, rather, was let into the case linings at its ends and supported along its length by blocks resting on the bottom, so that the air chamber below the soundboard was not divided.

The soundboard barring found on 18th-century instruments is extremely variable, some instruments having a series of diagonal ribs passing under the bridge approximately at right angles to its tenor section, while others have a cut-off bar running parallel to the tenor section of the bridge to separate the bridge from the triangular portion of the soundboard nearest to the keyboard. (In those instruments having octave strings in the bass, a cut-off bar of this sort serves as a hitch-pin rail for the octave strings.) Some instruments have a similar diagonal rib behind the bridge that cuts off the soundboard at the back right-hand corner as well. The use of one or even two decorated soundholes in the soundboard is characteristic of the work of such makers as Johann Heinrich Silbermann, Christian Gotthelf Hoffmann and Gottfried Joseph Horn.

Since clavichords were probably rarely used in conjunction with other instruments, their pitch is not of such interest as that of harpsichords and other accompanying instruments. Scalings of about 26–28 cm at *c''* were widely employed in the 18th century; these would have been suitable for the prevailing pitch when strung with brass wire. Other clavichords have scales which are both longer and shorter than these averages. It is not simple to infer the pitch from the scale of a clavichord, since part of the instrument makers' craft is to vary both the string size and the scale (and the length of wire from the tangent to the hitch-pin) so as to produce a suitable touch. Thus, some scales which are close to 26–28 cm were probably intended for the usual brass stringing. Others were almost certainly conceived for iron stringing, for example the 1775 Christian Gottlob Hubert clavichord (in the Musikinstrumenten-Museum, Karl-Marx-Universität, Leipzig), which has a *c''* of 31.2 cm. This would have given a brighter tone than brass wire. Although the 1725 Johann Conrad Speisegger clavichord (in the Musikinstrumenten-Museum, Leipzig) is built in the later style with a single bridge and diagonal stringing, its scaling is reminiscent of the 16th-century clavichords. It is possible that it was intended for iron stringing at a high pitch: this would certainly have improved the tone of the short bass strings.

5. SPAIN, PORTUGAL AND SCANDINAVIA

The interest in clavichords displayed by Bermudo and Santa María in the 16th century continued on the Iberian peninsula in the 17th and 18th centuries, although the instruments themselves appear to have been predictably conservative and intended for use principally as practice instruments. As late as 1723–4 Nassarre wrote about clavichords having a range of only *C/E* to *a''* and equipped with divided bridges, but he went on to mention smaller instruments having no more than two tangents striking the same pair of strings. In surviving Spanish and Portuguese instruments of this type, there tends to be a different arrangement of tangents from that found in German instruments fretted in pairs. Instead of the E's being paired with E♭ and the B's with B♭ so that the D's and A's are left unpaired, the Iberian instruments (and some Scandinavian ones) pair E♭ with D and B♭ with A, leaving the E's and B's unpaired.

The surviving Scandinavian clavichords, of which there are approximately 100, mostly suggest a domination of Scandinavian builders by German influence, especially that of the Hamburg school. Some of the late Scandinavian instruments do, however, include distinctive technical features, notably a means of guiding the keys at the back by covering their ends with leather and allowing each to slide in a wide channel, much like the keys of a Viennese piano. Certain Scandinavian clavichords have extended ranges only rarely seen in Germany in the 18th century, including *F'* to *a'''* in instruments by Pehr Lundborg of Stockholm and *F'* to *c''''* in instruments by Pehr Lindholm and G. C. Rackwitz, both also of Stockholm (a range exceeded in only one other surviving 18th-century example, a clavichord made in Spain with an extraordinary six-octave range, *F'* to *f''''*). The larger Scandinavian instruments, including all these wide-range ones, have an additional set of octave strings for the first one and a half octaves in the bass like those found in the clavichords of the Hamburg school.

6. THE MODERN REVIVAL

The revival of the clavichord, like that of so many other early instruments, was largely initiated by Arnold Dolmetsch in the late 1880s. In fact, it was stated in 1930 that the clavichord had been in continuous use in the Dolmetsch family since the early 19th century, so in this instance his work represented a survival of an instrument rather than its revival. As with much of Dolmetsch's other early work, his first clavichords were closely modelled on first-rate original instruments, in this case on large unfretted examples from the late 18th century. He later (reportedly at the suggestion of Violet Gordon Woodhouse, for whom he is said to have made no fewer than five clavichords) began to make far smaller instruments (still unfretted), which became a fairly standard model for other 20th-century builders. Many of these builders have been English, and between the wars a substantial number of German makers began building clavichords, followed by even more since World War II. In contrast to harpsichord making, clavichord making in the USA is marginal.

Like the harpsichord, the clavichord since the modern revival has been the subject of considerable experimentation in design and in the incorporation of modern materials. John Challis introduced a cast-aluminium frame and phenolic-resin wrest plank, and eventually an aluminium soundboard and bridge like those in his late harpsichords. Down-striking actions have been tried with some success despite the apparent drawback of sacrificing some of the velocity available with an ordinary balanced key-lever. A number of builders have revived variations on the early transversely strung design, generally in polygonal or spinet-shaped cases, in the hope of eliminating the twisting tendency of diagonal stringing in a rectangular case; especially in view of the fact that these instruments are unfretted and consequently have fairly wide stringbands, their keys are likely to vary enormously in length from treble to bass, which introduces certain problems in touch.

Although virtually all the modern clavichords made before the 1960s were unfretted, a number of makers have begun to build instruments modelled on the pairwise-fretted clavichords of Christian Gottlob Hubert and some of the 17th-century makers, as well as reconstructions based on 15th-century writings and pictures.

CHAPTER FOUR

Related Instruments

A number of instruments related to the harpsichord or clavichord families have some basic similarities, but warrant separate description. These instruments illustrate some of the by-ways and dead ends in instrument making, or, in the case of the chekker, the arpicordo and the arpitarrone (see §1–3 below), highlight the problem of understanding and interpreting archival material. The clavicytherium (see §4 below) is in essence an upright harpsichord, but since the keyboard and the strings are not in the same plane, a more complicated mechanism is necessary to operate the jacks. A combination instrument, the harpsichord-piano (see §5 below) was developed in the 18th century, when both the new fortepiano and the harpsichord were popular, but its design was complicated and few instruments were made. Another such instrument is the claviorgan (see §6 below), in which a harpsichord and organ are combined. In both these combination instruments the sound of the harpsichord remains unmodified and no separate repertory developed. An invention which led to an overlap between repertories was the lute-harpsichord (see §7 below), in which gut strings were plucked by the harpsichord mechanism so that the sound of music of the lute could be realized with less technical difficulty at the keyboard.

In the harpsichord, virginal and spinet, and related instruments, the strings are plucked. In the clavichord the strings are struck, and this method of sound production was used in various other keyboard instruments. One of the earliest was the dulce melos (see §8 below), briefly described by Arnaut de Zwolle (*c*1440), and other types were known: for example, the tangent piano (see §9 below), which owes something to both the harpsichord and the piano, having a sound somewhere between the two. Another way of exciting string motion is by bowing. One early bowed keyboard instrument was the *Geigenwerk*, invented by Hans Haiden (1536–1613), in which the strings were bowed continuously by parchment-covered wheels, an action similar to that of the hurdy-gurdy (see §10 below).

1. THE CHEKKER

'Chekker' is the earliest term used in archives and other writings to denote a string keyboard instrument; its exact meaning is still unclear and the subject of debate and research. A number of versions of the name occur: in French it is variously given as *archiquier*, *eschaquier*, *eschequier* or *eschiquier*, and the Spanish (Aragonese) names, *eschaquer* and *scaquer*, are closely related. Latin forms include *scacarum* and *scacatorium* and in German it occurs as *Schachtbret*. There appears to be no Italian equivalent. Recent research suggests that 'chekker' was simply the name given to the clavichord in the 14th and 15th centuries. Before the evidence for this identification is presented, some other suggestions that appeared in earlier writings should be mentioned.

Farmer (1926) suggested that the name is derived from the Arabic 'al-sẖaqira' and tentatively identified this as a virginal, but there is no supporting evidence. Some writers identified the chekker as an upright harpsichord (i.e. a clavicytherium), since a letter written by King John of Aragon in 1388 referred to 'an instrument seeming like organs, that sounds with strings', but the instrument was not named. Galpin ('Chekker', *Grove 4*, suppl.) believed that the 'dulce melos' described by Arnaut de Zwolle (*c*1440) was identical with the chekker. However, instruments with hammer action, such as the dulce melos, appear to have been rare, whereas the name 'chekker' appears frequently, and there is no evidence to support this identification. Galpin further suggested that the chekker's name was derived from the fact that the action was 'checked', in the sense that the motion of its keys was stopped by a fixed rail; this is unconvincing and could in any case apply to a clavichord, a harpsichord or a virginal. These suggestions can therefore be disregarded.

Documentary evidence reveals only that 'chekker' denoted a string keyboard instrument. Ripin (1975), who examined all the available documents, believed that he had found direct evidence for identifying the clavichord with the 'chekker': a sentence in a French court account book for 1488 refers to the purchase of 'un eschiquier ou manicordion'. However, 'manicordion' could be used in two ways: it was applied generally to any string keyboard instrument and specifically to the clavichord. Thus, 'ou manicordion' could have been added to provide a more precise explanation of the instrument (as Ripin believed), or the person responsible for the accounts might not have known how to describe it and therefore covered the possibilities by giving the two names. In examining such sources it is usually possible to find more than one plausible explanation.

While the evidence shows that 'chekker' referred to a string keyboard instrument, it does not indicate the type of action it had; it is necessary therefore to pursue other lines of inquiry. Meeùs (1985) has argued that since 'chekker' is the earliest known name for a string keyboard instrument, the earliest such instrument must be the one denoted by that name. Provided that it is the name for a specific instrument and that the earliest known string keyboard instruments (i.e. clavichord, harpsichord, virginal and clavicytherium) did not come into being at the same time, this argument is correct. The chronology of documents about the clavichord, harpsichord, virginal and clavicytherium suggests that they appeared in the order listed here. There is evidence for the clavichord around the middle of the 14th century and the earliest reference to the chekker is 1360. Evidence for the harpsichord appears towards the end of the century (*c*1390); Arnaut's treatise of around 1440 provides the first reference to an instrument identifiable as the virginal; and the first clavicytherium reference is around 1460. Although the earliest references to the harpsichord occur soon after the earliest references to the clavichord – and it may be argued that there can therefore be no certainty that the clavichord was the earlier instrument – the harpsichord is wing-shaped, and it is probable that the chekker, like the clavichord, was rectangular (see below). (The rectangular virginal is not documented until almost a century after the clavichord.)

A second argument was initiated by Ripin (1975), who suggested that the name 'chekker' was derived from 'exchequer', the medieval counting-board. Meeùs (1985) has made a persuasive case for this derivation. The title-page of Gregor Reisch's *Margarita philosophica* (1503) shows a lady operating a counting frame (an exchequer) and she appears at first glance to be playing a clavichord. The four or five lines on the rectangular board give it a sufficiently close resemblance to a clavichord for it to be conceivable that the name 'exchequer' might have been applied to a keyboard instrument. Since the exchequer would also have been used for the mathematical computations of music theorists it would have been well known in musical circles. Besides this practical convergence of music and mathematics, it should be remembered that the two disciplines were fundamental to the liberal arts. Mersenne (1648) used the Latin 'abacus' (the modern English word for 'exchequer') for keyboard and this seems to confirm the natural alliance of the idea of the exchequer and a keyboard instrument. The probability that 'chekker' is derived from 'exchequer' suggests that the chekker was rectangular. Moreover, if the chekker was the earliest string keyboard instrument, then an early rectangular instrument must be identified with the clavichord, a conclusion previously reached only from chronological evidence.

The derivation from 'exchequer' can be employed also to show which of the rectangular instruments in use at the time might have been called 'chekker'. Of the 26 known 15th-century representations of rectangular instruments, at least 20 are probably of clavichords (it is not always possible to identify the instrument with certainty, as parts of the action or stringband are not visible). The existence of the clavichord in the 15th century is well documented, both as a musical instrument and as a theoretical example of the proportioning of string lengths for chosen intervals. Of the virginal at that time relatively little is known; Virdung (1511) stated that it was made in a rectangular case like a clavichord. The virginal was apparently known to Arnaut de Zwolle, although he gave no name for the instrument he described briefly; a fragment inserted in his manuscript (including a sketch) mentions a rectangular instrument with plucking action ('monochord sounding as loudly as a harpsichord'). Since the clavichord was much better documented, it is likely that it was more common than the virginal and therefore more likely to be the chekker.

Page (1979) has argued that it is a fundamental misconception to suppose that 'chekker' denoted a particular instrument, since terminology at that time was likely to have been imprecise. He inferred that Arnaut's 'clavichordium' could denote tangent, hammer or plucking action. Meeùs (1985), however, has argued convincingly that Arnaut used the name 'clavichordium' in a didactic or illustrative way: thus, when he wrote that the clavichord could be 'transformed' into a dulce melos or virginal by changing the action, it must be understood that he was not proposing an actual modification, because that would have been technically impossible. The clavichord was the instrument named (with other instruments in terms of it) simply because it was familiar to the reader. Galilei (1581) used the same type of analogy when he described the stringing of an unusual harp by comparing it to a 'gravicembalo' (i.e. a harpsichord): the *gravicembalo* was familiar to his readers, but the harp was not. Thus it can be inferred that if the clavichord was used to describe another instrument, it was more familiar than the other instrument.

While Page's argument is not convincing, the question of whether 'chekker' was used in a general or imprecise way must be asked. There are many examples of such use in the history of musical instruments (i.e. 'manicordion' above). It is even more important to establish whether the name had a general sense from the beginning. It is feasible that in deriving the name from the exchequer the intention was to refer to a part of an instrument, such as the keyboard (similar to Mersenne's use of 'abacus'). In that case the portative organ (i.e. a wind keyboard instrument) might also have been called 'chekker', except that in some documents the chekker is clearly a string keyboard instrument. When considering that 'chekker' might have denoted any rectangular string keyboard instrument it should be remembered that, of the instruments recorded, evidence shows that the clavichord was in use long before the virginal and the dulce melos is known only from Arnaut's description. In view of the paucity of information unshakable conclusions cannot be reached. Nevertheless, 'chekker' seems to have denoted a specific instrument and evidence points clearly to the clavichord.

2. THE ARPICORDO

The arpicordo (sometimes called *harpicordo* or *ampicordo*) has not attracted as much attention as the chekker, although there has been a good deal of confusion regarding its identity. The term appears only in Italian, and despite its similarity to the English 'harpsichord', it does not denote that instrument; the modern Italian terms for the harpsichord are *clavicembalo* and *cembalo*, and earlier terms were similar (e.g. *clavazimbalo*, *clavacinbalo*). 'Arpicordo' is sometimes also confused with the *arpichordum* stop used on some Flemish virginals, where small metal pins are brought into contact with the vibrating strings so that a buzzing sound results. Cervelli (1958, 1973) has shown that the arpicordo can be identified with the polygonal (usually five- or six-sided) virginal (or 'spinetta') made widely in Italy in the 16th and 17th centuries. Several sources, including Galilei (1581), mention the arpicordo, and Adriano Banchieri (2/1611) attempted to describe its derivation from the harp. His description is awkward (and has led to confusion), but his intention was to show that the shape of a harp is reflected in the bridges of the arpicordo. While it is true that the shape of the harp is found in the layout of the bridges, this shape is also found in the rectangular virginal, which Banchieri called a 'spinetta'. His description falls short of being a definition of an arpicordo: the shape of the layout of the bridges is just one necessary condition. Since the name 'arpicordo' came into use long before Banchieri was born, he cannot be a first-hand expert on the origin of the instrument. However, he seems to have felt an obligation as a writer to provide definitions for his readers. His explanation that the name 'spinetta' came from Giovanni Spinetti (according to Banchieri, a Venetian maker who invented the instrument) is not convincing and reinforces the impression that Banchieri wished to appear authoritative. His comparison of the arpicordo to the harp led to the erroneous speculation (Neven, 1970) that the arpicordo had some mechanism to produce a buzzing sound, as in some harps; in effect, an *arpichordum* stop. If this were the case, it would be expected that such a mechanism would be found among the large number of surviving 16th-century arpicordi. Since these instruments have recently been thoroughly researched and no such examples are known, it is obvious that an *arpichordum* stop is not a characteristic of the arpicordo.

Some makers of arpicordi have been identified: Lanfranco (1533) described Giovanni Francesco Antegnati of Brescia (*fl* 1533–44) as a maker of 'monochordi [clavichords], Arpicordi, & Clavacymbali'. No clavichords or harpsichords made by Antegnati are known, but four arpicordi have survived.

The term 'arpicordo' was used from at least the beginning of the 16th century, but seems to have gone out of favour towards the middle of the 17th. Thereafter the arpicordo is usually referred to as a 'spinetta', sometimes as a 'spinetto'.

3. THE ARPITARRONE

Banchieri (2/1611) claimed to have invented a new instrument which he named 'arpitarrone'. The name is derived from two other instruments, the arpicordo and the chitarrone. 'Arpicordo' (see §2 above) was the 16th-century name for a polygonal virginal (sometimes also a polygonal spinet; see Chapter Two, §2), but the chitarrone was in effect a bass lute. Its long neck permitted string lengths of 140–170 cm which imparted a powerful character to the bass register. Although some instruments were wire strung, the majority probably had gut strings. Since an arpicordo was in any case wire strung, it would be an advantage to combine the two instruments only if each brought something different to the union. Thus, the arpitarrone was probably a gut-strung keyboard instrument and thereby a precursor of the lute-harpsichord (see §7 below), in effect, a lute-virginal. Michel de Hodes, an instrument maker in Milan, had made an 'arpicordo leutato', that is, a gut-strung arpicordo, but Banchieri's contribution was to remove some notes from the compass in the treble and add some in the bass so that the instrument had a deeper range. The arpitarrone had a compass of *C*,*D* (i.e. no *C*♯) to *e*″.

4. THE CLAVICYTHERIUM

The clavicytherium (French *clavecin verticale*, German *Klavizi-terium*, Italian *cembalo verticale*) is an upright harpsichord, that is, one in which the soundboard is vertical rather than horizontal. Its sound projects more directly into the room, so that the clavicytherium appears louder than an ordinary harpsichord. Some writers, such as Praetorius (2/1619), regarded the clavicy-therium as no more than a vertical harpsichord. Others (see Bonanni, 1722, and Adlung, 1768) regarded it as distinct from a harpsichord, emphasizing the player's different perception of the sound as evidence of the difference between the instruments.

48. Clavicytherium, c1480

Since the jacks have to move horizontally, a complex mechanism is required to return them to their rest position. No standard mechanism evolved, but in most instruments springs are necessary. This often results in a fairly heavy touch and unresponsive action. Two basic forms of the clavicytherium are known. These concern the shape of the soundboard: the instrument may resemble a harpsichord as viewed from above (from the keyboard end), with a single bentside at the right-hand side (see fig.48), or a symmetrical, 'pyramid', shape is produced with two bentsides, left and right. With the pyramid shape other difficulties arise, since the longer, bass strings are in the middle and the treble strings are at the sides, and a system of intermediate levers (rollerboard) is required to permit each key to play the correct string unless diagonal stringing is used.

The earliest known reference to a clavicytherium is in the manuscript treatise of Paulus Paulirinus of Prague (*c*1460), but the description is vague. Sebastian Virdung, in his *Musica getutscht* (1511), stated that the instrument had been newly invented. According to Virdung the clavicytherium had gut strings and it would therefore have sounded like a harp rather than a harpsichord (which has metal strings). The surviving clavicytheria appear to have been made for metal strings, so it is unclear whether many were made as described by Virdung.

Since both Virdung and Paulus Paulirinus indicate that the instrument was invented in the 15th century (and if Virdung is right, in the second half of the 15th century), it is somewhat surprising to find that the oldest known clavicytherium can be dated to about 1480, shortly after its apparent invention. This clavicytherium (in the museum of the Royal College of Music, London; fig.48) is thought to be the earliest surviving string keyboard instrument, being about 35 years older than the oldest known harpsichord (1515; see Chapter One, p.15) and was probably made in Ulm. It is 142.5 cm high (including a stand of 6.9 cm). It differs from the usual harpsichord shape in that the bentside meets the spine at a point (i.e. there is no intervening small tailpiece). It has a unique and simple action in which the key, a vertical lever and the forward-projecting jack are all assembled into a single rigid piece. When the key is depressed the entire assembly rocks forward so that the jack (moving along the path of an arc) is forced past its string; when the key is released, the assembly falls back again under its own weight, returning the jack to its original position.

Although the keyboard now has the range *C/E* to g″ it is clear that this has resulted from a modification: the original range was *E*, *E*♯, *F*, *G* to *g*″ (see Debenham, 1978). Thus, the first accidental in the bass octave was *G*♯; the apparent *E*♯ cannot be an actual *E*♯, as such an interval would have a place only in an enharmonic keyboard. The problem is, then, to establish how the lowest two notes (*E* and *E*♯) were tuned, and possibly also the *G*♯, since this may have sounded another note. In any event, this clavicytherium testifies to the use of notes below *F* at this early date. Bartolomeo Ramos de Pareia (1482) mentioned that he had encountered in Bologna a string keyboard instrument (though he did not specify what sort) that reached to *D*. It may well be that the first two notes of the clavicytherium compass were *D* and *E*, the *E* being played from the apparent *E*♯ and the *E* key sounding as *D*. This arrangement, where an apparent sharp plays a natural note, is of course found in the *C* short octave, which was widely used in the 16th century. Virdung (1511) stated that some instruments included *F* and *G*♯ and Arnolt Schlick (*Spiegel der Orgelmacher und Organisten*, 1511) preferred to have an *F*♯ and *G*♯ even though this arrangement was not often provided on organs. If the apparent *G*♯ in the clavicytherium was not a *G*♯, there would be enough keys in the bottom octave to provide a full diatonic octave, that is, *C*,*D*,*E*,*F*,*G*,*A*, *B*♭. However, in view of the early date of this instrument the compass starting on *C* is less likely. It is also possible that the lowest notes had no intended pitch, but were tuned as required.

Although some clavicytheria are identical to harpsichords in the length of the soundboard, the scaling of the strings and in their pitch, this early German instrument is relatively small. Thus, the question arises whether this instrument is the same size as late 15th-century harpsichords (of which no examples are known), or whether it is especially small for other reasons. The scale is short: *c*″ = 22.5 cm (24 cm when calculated from the most highly stressed (*g*″) string); this suggests a pitch higher than normal. It is interesting that the pitch suggested by this instrument is virtually identical with that of the harpsichord drawn by Arnaut de Zwolle (*c*1440). Although he gave no measurements, it is possible to infer a range of scalings by extrapolation from a likely range of keyboard widths. This procedure shows the close similarity of the two instruments. Since a number of 15th-century illustrations depict relatively small harpsichords, it seems no accident that these two instruments are of similar size. Furthermore, the scalings of these instruments would make them about a 5th higher in pitch than the harpsichords built in the 16th century having scales of about 35 cm at *c''* (assuming that all instruments were strung with the same material). There is little evidence from which conclusions can be drawn, but this difference in pitch of a 5th may have been no accident.

Another early instrument (perhaps from around 1600) is in the Norsk Folkemuseum, Oslo. Several Italian clavicytheria purporting to date from the 16th and 17th centuries are forgeries in whole or in part. Some have quite obviously been constructed from parts of harpsichords.

Virdung was the earliest writer to use the term clavicytherium for the upright harpsichord. His crude woodcut illustration (which has been reversed by the printer) shows a 38-note range, *F*,*G* to *g''* (i.e. lacking *F*♯). Praetorius showed an example (2/1619; see fig.43) and wrote that its sound was like that of a cittern or harp. As he likens citterns to harpsichords in their sound, but describes both metal- and gut-strung harps, it is not clear whether Praetorius's clavicytherium was strung with gut or metal. The majority verdict would appear to be in favour of wire.

A German instrument of Praetorius's period (in the Germanisches Nationalmuseum, Nuremberg) has two sets of unison strings at normal pitch and a third set tuned an octave higher (2 × 8′, 1 × 4′). There are four rows of jacks, one of which plucks one of the sets of unison strings very close to the nut to produce a penetrating, nasal tone; in addition, a sliding batten with leather pads can be moved to mute the sound of one of the unison strings. In this instrument a vertical arm set in the back of each key has four finger-like projections, each of which fits into a slot in one of the four jacks provided for each note. Since the connection between the jacks and this arm is not a fixed one, it is possible for the jacks to move forwards and backwards horizontally instead of in a curved path. It is also possible to change registers by shifting the jackslides as in a harpsichord.

An Italian clavicytherium of symmetrical shape, possibly by Cristofori, is preserved in the Museo degli Strumenti Musicali in Rome. It has a compass of *C*/*E* to *c'''*, and has two unison strings for each note. Martin Kaiser, who worked for some time in Venice, made a clavicytherium in about 1675 with a similar symmetrical shape. It was formerly owned by the Emperor Leopold I and is now in the Kunsthistorisches Museum, Vienna.

In the late 17th century and in the 18th century, clavicytheria were built throughout Europe, including Scandinavia and Great Britain. Cristofori made one in 1697 which is recorded in a Medici inventory of 1700, and an enormous example about 365 cm high was constructed by Nicolas Brelin in Sweden in 1741; it had eight register-changing pedals. A number of handsome clavicytheria were made in Dublin in the second half of the 18th century; one by Ferdinand Weber uses oblique stringing to achieve a symmetrical 'pyramid' form. Three important examples by Albert Delin (who is known to have worked in Tournai between 1750 and 1770), are notable for their smoothness of action and fine tone. Delin's action uses a pivoted bell-crank, the horizontal arm of which is pushed upwards by a vertical sticker resting on the back of the key; the jacks are hooked into the vertical arm of the bell-crank, which is so balanced that when the key is released it brings the jacks back to the rest position without additional weights or springs.

5. THE HARPSICHORD-PIANO

Towards the end of the 18th century, as the popularity of the harpsichord declined and that of the fortepiano increased, some makers attempted to combine the harpsichord and fortepiano in one case. These instruments were made either in the conventional shape with some means of changing from one mechanism to the other, or like the so-called 'vis-à-vis' harpsichord-piano, which was basically two separate instruments in one rectangular case with the keyboards at opposite ends.

The earliest experiments must predate 1716, when Jean Marius submitted models of four different 'hammer harpsichords' (*clavecins à maillets*) to the Académie Royale des Sciences in Paris, one of which was a combined piano and harpsichord with both hammers and jacks. An 'arpicembalo' by Cristofori (listed in a Medici inventory, apparently from 1700) also had hammers and a jack mechanism. In the combination instrument demonstrated by Weltman before the Académie Royale des Sciences in 1759, however, the sounds generated by the plucking action were contrasted with piano-like tones produced by a mechanism anticipating that of the tangent piano (see §9 below), rather than by the normal hammer action used by other makers of the harpsichord-piano.

Combination instruments were produced by some of the most distinguished early piano makers, notably Johann Andreas Stein of Augsburg, Sébastien Erard of Paris, Robert Stodart of London, and Giovanni Ferrini. Ferrini, who worked for Cristofori in Florence, built a combined harpsichord-piano in 1746 (now in private ownership in Bologna; see Tagliavini and Van der Meer, 1986). It has two keyboards at one end; one for the harpsichord action, the other for the piano, but only one set of strings. Stein in particular was captivated by the notion of the harpsichord-piano and persisted in building a variety of models over a number of years. In 1769 he brought out his 'Polytoniclavichordium', which despite its name was a combination of a two-manual harpsichord (disposed 1 x 16′, 3 x 8′) and a piano, each with its own set of strings to permit string scaling and striking points appropriate to each type of instrument. By 1777 he had followed this with the *vis-à-vis* harpsichord-piano. The two surviving examples of this instrument, in Verona and Naples, have three keyboards at one end (a two-manual harpsichord plus piano) and a single keyboard for a second piano at the other end. The earlier (Verona) instrument includes in its harpsichord portion a 16′ stop in addition to the normal two unisons and 4′. The slightly less elaborate instrument in Naples was considered a worthy gift from Emperor Joseph II of Austria to King Ferdinand when presented in 1784 in appreciation of the king's hospitality. For those in more modest circumstances Stein created the 'Saitenharmonika' in 1788, described as a 'bichord pianoforte' with a special 'spinet stop' allowing the instrument to be played as either a piano or a spinet. It is not known whether this simpler device was a true harpsichord-piano or merely a piano fitted with one of the then prevalent special-effect stops. At least one of Stein's apprentices, Ignaz Joseph Senft of Koblenz, is known to have constructed a *vis-à-vis* harpsichord-piano in 1793.

No example of Erard's 'clavecin mécanique' of the 1770s is known, but the instrument is said to have been a combined harpsichord and paino with separate keyboards that could be coupled. Robert Stodart's 1777 patent for a combination instrument (including a harpsichord disposed 1 x 8′, 1 x 4′) gives the first detailed drawing of the English grand piano action. (The harpsichord-piano in Washington signed 'Robertus Stodard et Co.' is now thought probably to be the work of a later English maker, James Davis, who obtained a similar patent in 1792.)

John Joseph Merlin was an instrument maker in London whose inventiveness ran to such extremes as the 'Barrel Harpsichord', the 'Patent Piano Forte Harpsichord with Kettle Drums' and the 'Patent Double Bass Piano Forte Harpsichord', in addition to such non-musical inventions as the roller-skate and wheel-chair. He patented a compound harpsichord in 1774. An instrument of his dated 1780 (in the Deutsches Museum, Munich) is remarkable as the only surviving English harpsichord that includes a 16′ stop as well as the normal 8′ and 4′. The downstriking piano hammer action sounds both the strings of the 8′ harpsichord register and a second set. In addition the instrument is fitted with a device for recording improvisations by mechanically inscribing pencil lines on a moving strip of paper. While many harpsichords in the late 18th century were subsequently converted to pianos, Merlin is the only maker known to have converted them into combination instruments.

The harpsichord-piano seems never to have gained wide popularity. No music demanding its special capabilities is known, and by the 1790s the instrument was seemingly extinct. However, in 1861 a patent was issued for an invention by Robert Thomas Worton, the 'lyro-pianoforte', which combines both jack and hammer actions for plucking and striking the strings.

6. THE CLAVIORGAN

'Claviorgan' is the English equivalent of the quasi-Latin *claviorganum*, denoting a keyboard instrument in which strings and pipes 'sound together to produce a pleasing sound' (as described by Praetorius, 2/1619). The French term is *clavecin organisé*, the German *Orgelklavier*, the Italian *claviorgano*, and the Spanish *claviórgano*. In early sources, late 15th-century Spanish or 16th-century Italian collections, for example, it cannot be assumed that *clabiórgano* or *claviorgano*, etc, invariably denote a composite keyboard instrument of this kind; often the word may have been used for secular organs in general, perhaps to distinguish them from portatives or regals. The English term seems to have appeared only at the end of the 19th century (when it was used by such authors as Engel (1874) and Hipkins (1896)) but it may have been used earlier. The adjective 'organisé' was used in France at least by the middle of the 16th century and copied by English lexicographers such as Cotgrave (1611).

The true claviorgan remained on the fringe of music-making for at least three centuries; its history is thus neither continuous nor connected, but comprises a series of important types. In the 16th century, spinets or virginals 'with pipes undernethe' are known to have existed from documentary evidence (e.g. at least five are listed in the inventories of Henry VIII, 1547) and from surviving examples (e.g. the spinet-regal-organ formerly in Schloss Ambras); double- or triple-strung, full-size harpsichords with positive organs incorporated are to be found in Germany (a Dresden inventory, 1593), Italy (Banchieri, 1605), England (one made by Theewes, 1579) and elsewhere. Many examples must have been little more than toys (e.g. the mechanical organ patented in Venice, 1575). In the 17th century there was an immense variety, from the clavichord-organ combinations known from theorists (Todini, 1676) and extant examples (V. Zeiss of Linz, 1639) to the organ-spinet-harpsichord-Geigenwerk made in Rome by Michele Todini in about 1670. The acoustic and mechanical theorists (e.g. Kircher, 1650) were attracted to the more doubtful aspects of these composite instruments. From about 1580 to about 1780, many large organs are known to have had a row or two of harpsichord strings, especially those in German court churches, but also in various other places from Sicily to Coventry. During the 18th century, particularly in England, chest-like or even harpsichord-shaped chamber organs were made specifically to carry on top a harpsichord whose keys depressed the organ pallets below by means of simple stickers; this is known both from theoretical sources (e.g. the varied and detailed drawings given by Dom Bédos, 1766–78) and from extant examples (e.g. the Earl of Wemyss's Kirckman–Snetzler claviorgan of 1745–51; see fig.49). Even in many quite late sources it is not clear what exactly 'claviorgano' denoted (e.g. Cristofori's accounts, Florence, 1693). Late in the 18th century many pianos, particularly large square ones, were made with several ranks of flute and chorus organ pipes, often by the best makers in London (e.g. Broadwood) and to a lesser extent in Paris, where Taskin was active in this field. By about 1840, harmonium-pianos

49. Claviorgan by Jacob Kirckman and John Snetzler, 1745–51

played only a minor role amongst the vast array of composite, hybrid and other fanciful, constantly patented inventions.

There never was a specific claviorgan repertory, although some pieces were written to exploit individual instruments: for example, one of Handel's organ concertos of 1739 was composed for the organ-harpsichord used in *Saul* (1738). Claviorgans were used occasionally in the Florentine *intermedii* in the 16th century. Many theorists pointed out the varieties of colours possible, often calculating them mathematically. Claviorgans were occasionally named on the title-pages or in prefaces to various publications (e.g. A. de Arena, *Bassas dansas*, 1572, for 'espineta sola, espinate organisati', and S. Seminiati, *Salmi*,

1620, for 'leuti . . . organi o claviorgani'). They are sometimes mentioned in diaries and the like as having been played in works not expressly calling for them. For example, Michael Arne played a theatre concerto in 1784 on an 'Organized Piano Forte' (R. J. S. Stevens's MS, in the Pendlebury Library of Music, Cambridge); Burney (1771) heard an Italian nun using a claviorgan in church; Mattheson (1739) recommended them in Hamburg for church cantatas. By about 1770 a *clavecin organisé* was played in order to give dynamic changes by adding or subtracting organ stops. In 1768 Adlung wrote that the claviorgan was less common than in his 'young days', the piano having replaced it in expressive music. A *clavecin organisé* made by J. A. Stein is in the Historiska Museum, Göteborg.

7. THE LUTE-HARPSICHORD

The lute-harpsichord was a gut-strung harpsichord intended to imitate the sound of the lute. Other terms for it include *clavecin-luth* (French) and the German *Lautenklavecimbel*, *Lautenklavier* and *Lautenwerck*. It should not be confused with the so-called 'lute stop' on some harpsichords which is a row of jacks that pluck one of the unison registers close to the nut: clearly, the sound of the close-plucked metal strings of the ordinary harpsichord is not the same as that of plucked gut strings. The buff stop found on many harpsichords has also been given various names which indicate that its sound was thought to be like that of a lute (e.g. 'jeu de luth' and 'liuto'), but it is a device that partly damps the vibrations of metal strings. Some writers have described the arpicordo as a form of lute-harpsichord, but this is incorrect because 'arpicordo' is only another name for the Italian polygonal virginal (see §2 above). Gut-strung arpicordi were known, however – Michel de Hodes made an 'arpicordo leutato', and Banchieri's arpitarrone was probably gut-strung – and in all probability there were other experiments with gut-strung keyboard instruments. A 'Harfentive' was described by Virdung (1511) as being gut strung, but exactly what kind of instrument it was is unclear.

German makers in the first half of the 18th century seem to have been those most interested in the potentials of the lute-harpsichord and a number of different types were produced by such builders as Johann Christoph Fleischer, Zacharias Hildebrandt and Johann Nicolaus Bach. The form and layout of lute-harpsichords was quite variable. Some were rectangular, some oval, some wing-shaped like a harpsichord; some had a hemispherical resonator below the soundboard (similar to the lute) and others had continuous bridges like those in a conventional harpsichord. Of all these instruments, Fleischer's 'Theorbenflügel' was probably the most elaborate, having three sets of strings: the register at 8′ pitch and the one tuned an octave lower were of gut, but there was also a 4′ register with metal strings, presumably to brighten the overall sound.

Jacob Adlung devoted a chapter of his *Musica mechanica organoedi* (ii, 1768, pp.133ff) to lute-harpsichords and considered them to be 'the most beautiful of all keyboard instruments after the organ . . . because it imitates the lute, not only in tone quality, but also in compass and delicacy'. He gives the compass as generally three octaves, *C* to *c*″, with strings that are not as long in the bass as in a harpsichord. The two lower octaves have two strings to every note and in the bass octave these are tuned as unison and octave, as on the lute; the top octave is single strung. According to Adlung the 'Lautenwerk' sounded so like the lute that it could deceive even experienced lutenists, but had the serious disadvantage of not being able to imitate the lute's dynamic gradations. J. N. Bach (a second cousin of J. S. Bach) partly overcame this difficulty by devising instruments with two or three keyboards. The jacks plucked the strings at different distances from the nut, those furthest from the nut giving the softest tone.

Among the instruments in the inventory of J. S. Bach's estate, made after his death in Leipzig in 1750, there were two lute-harpsichords, valued at 30 Reichsthaler each. An interesting eyewitness account of a lute-harpsichord which Bach is said to have designed, and had built for him by Zacharias Hildebrandt, is given by Johann Friedrich Agricola (1720–74), who was himself a pupil of Bach. Agricola wrote in Adlung's *Musica mechanica organoedi* (p.139) that

> It had two courses of gut strings, and a so-called Little Octave of brass strings. In its normal disposition – that is, when only one stop was drawn – it sounded more like a theorbo than a lute, but if one drew the lute stop [i.e. the buff stop] such as is found on a harpsichord together with the cornet stop [i.e. the 4′ brass strings], one could almost deceive even professional lutenists.

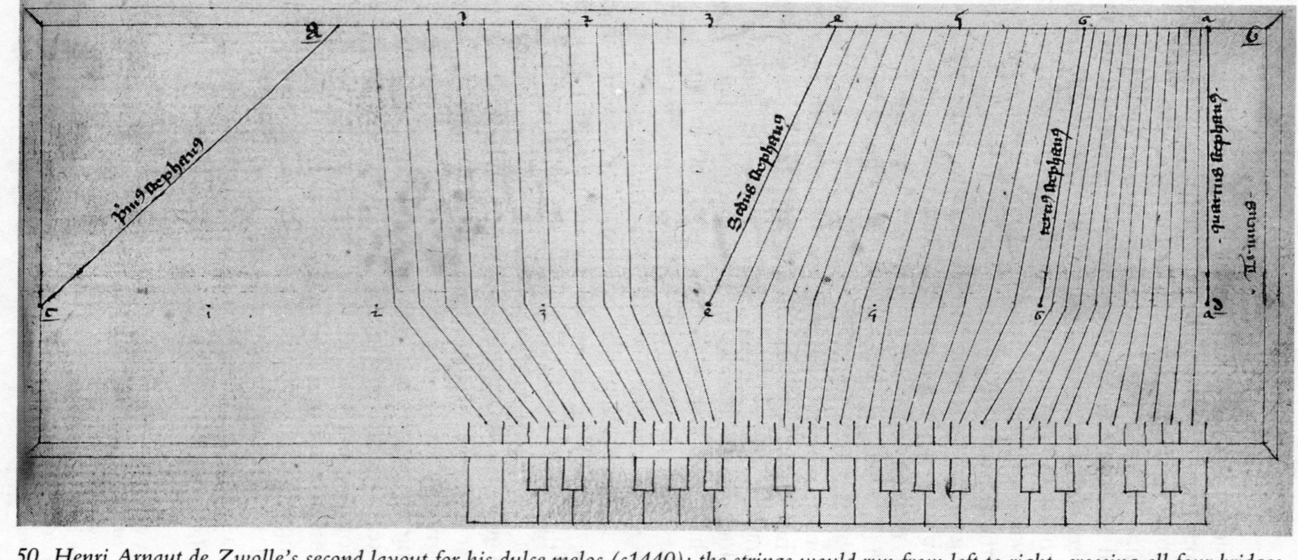

50. Henri Arnaut de Zwolle's second layout for his dulce melos (c1440); the strings would run from left to right, crossing all four bridges

One work by J. S. Bach that must surely have been written for a lute-harpsichord is the Suite in E minor, BWV996. Bach's autograph has not survived but a contemporary manuscript copy, by Johann Ludwig Krebs, has the following inscription on the title-page: 'Preludio con la Svite / da / Gio: Bast. Bach./ aufs Lauten Werck'.

8. THE DULCE MELOS

The term 'dulce melos', the Latin name for a dulcimer, owes its place in the history of keyboard instruments to the manuscript treatise by Arnaut de Zwolle (*c*1440; Bibliothèque Nationale, Paris, MS lat.7295). Arnaut used the term for a keyboard instrument that was essentially a keyed dulcimer whose action to a large extent appears to prefigure that of certain early pianos. The layout and stringing of the instrument resemble those of many dulcimers, since the sections of a string on opposite sides of the bridges are used to sound two different notes.

In Arnaut's instrument there were 12 unison pairs of strings passing over four bridges. The distances between the first and second bridges (counting from the left), the second and third, and the third and fourth were in the ratio 4:2:1; hence that section of any pair of strings struck between the second and third bridges would sound an octave higher than the section between the first and second; that between the third and fourth would sound an octave higher still. Each of the 12 pairs of strings was thus capable of sounding three notes: its basic pitch (struck between the first and second bridges) and that note one or two octaves higher. In this way the 12 pairs of strings, being tuned consecutively by semitones, provided a fully chromatic range of two octaves and a 7th, from *B* to *a''*, all the B's being sounded from the first pair of strings, all the C's from the second pair, and so on.

Arnaut gave two different layouts for his dulce melos. In the first, the bridges are all placed parallel to one another and perpendicular to the front of the instrument's rectangular case. All the strings thus have the same length, even though the highest and lowest pairs are tuned nearly an octave apart. In the second layout (see fig.50) the bridges are placed obliquely in order to mitigate this problem, but the difference in length between the lowest and highest strings is still far less than it should be for just scaling, and the keys are of necessity more cranked than they are in the first design. In both designs it would appear that the soundboard was near the bottom of the case, with the keys and action above it, the strings being carried on very tall bridges resting on the soundboard and rising between the groups of keys for each octave of the instrument's range.

The action that Arnaut specifically suggested for the dulce melos is the fourth of those very crudely depicted and cryptically described on the page of his manuscript devoted to the harpsichord (see fig.2 above, upper right-hand corner). The diagram and description are far from clear, but it seems that a staple-shaped metal 'hammer', carried by a slip of wood hinged to the key at the back, was thrown against the strings when the upward motion of the back of the key was stopped by a fixed obstacle. Although this mechanism has no exact parallel in any surviving piano action, principally because it includes no arrangement to make the hammer move faster than the key, it seems to have been close to the action seen in some mid-18th-century German square pianos. Here, as in all instruments with a German action, the hammers are mounted on the keys. There is no documentary evidence for Galpin's suggestion (in 'Chekker', *Grove 4*, suppl.) that the chekker employed the action of Arnaut's dulce melos.

No examples of a dulce melos are known and no other literary references have been found. One instrument which might have had an action similar to that of a dulce melos is the 'Instrumento Pian e Forte' built by Hippolito Cricca around the end of the 16th century in Ferrara. In a letter dated 31 December 1598, he requested materials for the construction of a third such instrument. It can be inferred that the instrument had a cypress soundboard and was strung with brass wire, but there are no clues as to the mechanism. A suggestion has been made (Boalch, 2/1974, p.29) that this was a harpsichord with two registers, with movable jackslides. Since Venetian harpsichords with two registers had been built since the 1530s with movable slides it would hardly have been much of a novelty in 1598, worthy of the description 'Pian e Forte'. There is, therefore, the possibility that this instrument had a striking mechanism. It was also coupled to an organ below it.

An interesting speculation is that Cristofori might have heard of this instrument, or even have seen it, and might have been stimulated to make his experiments with his piano escapement. The idea is by no means far fetched, since some instruments from the Ferrara court, where Cricca was employed, later found their way to Florence where Cristofori worked on his pianos.

9. THE TANGENT PIANO

The tangent piano comes between the harpsichord and fortepiano in the history of the manufacture of keyboard instruments. Its basic mechanism resembles that of the piano but instead of hammers it has small strips of wood (like jacks) that are propelled towards the strings. Thus the sound is capable of dynamic nuance and, depending on how the striking surfaces of the tangents are treated, can be bright as in a harpsichord, or mellower as in a fortepiano.

The most important instrument of this type was the 'Tangentenflügel' said to have been invented in 1751 by Franz Jakob Späth the younger and made by him (in partnership with his son-in-law Christoph Friedrich Schmahl after 1774) in Regensburg. The tangent piano principle was, however, incorporated in a number of other designs, both earlier and later, none of which can be shown to have had any direct connection with Späth's instrument. The same principle is embodied in the actions devised by Jean Marius in 1716 and Christoph Gottlieb Schröter in 1739 but not published until 1763. A grand piano action patented in England in 1787 by Humphrey Walton altered the ordinary square piano action by making the hammer propel a padded jack-like part which struck the strings. In addition, a number of harpsichords and virginals were converted in the 18th century to instruments of the tangent piano type simply by replacing their jacks with shorter slips of wood so arranged that they struck the strings instead of moving past them.

None of these converted instruments, however, includes the refinements found in Späth and Schmahl's *Tangentenflügel*, of which all the surviving examples seem to have been made after 1790. The action of these instruments includes an intermediate lever to increase the velocity with which the jack-like striking element is propelled towards the strings (see fig.51, p.194), as well as a large assortment of tone-modifying devices. These included means for raising the dampers, for introducing a strip

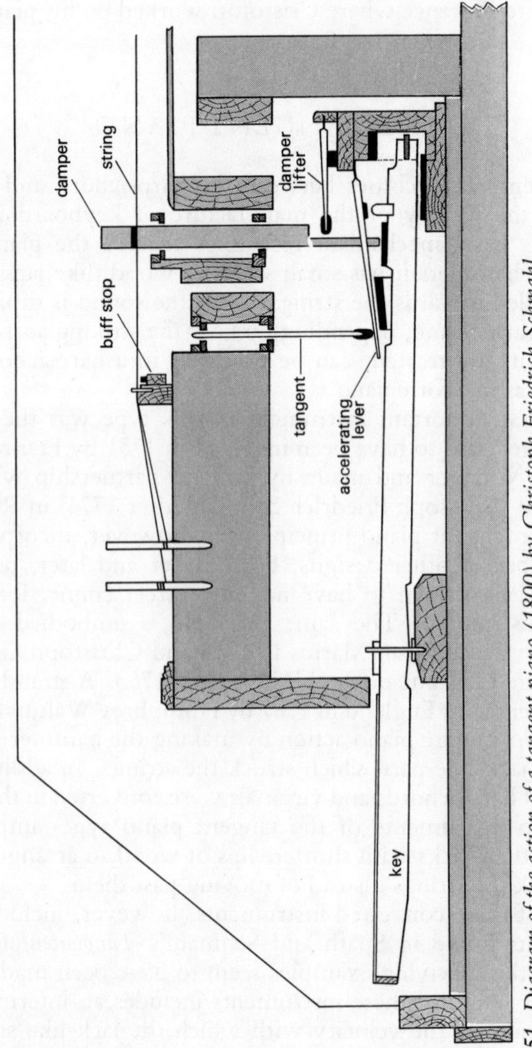

51. Diagram of the action of a tangent piano (1800) by Christoph Friedrich Schmahl

of cloth between the striking tangents and strings (similar to the moderator in the fortepiano), an 'una corda' stop whereby only one of the pair of strings is excited, and a buff stop that mutes the strings by pressing a piece of leather or cloth against them at the nut; moreover, in several of the surviving examples one or more of these devices can be used separately in the treble and bass. These instruments look very like grand pianos of the period, and, as in pianos, the loudness of the sound is determined by the force with which the keys are depressed, although the action is far less complicated.

The conversions from quilled instruments must be thought of as makeshifts and the tangent piano actions of Marius and Schröter were experimental constructions, each employed in only a single instrument (if indeed any instruments using these actions were ever built), but the developed tangent piano is neither an experiment, nor a compromise. Several instruments that have survived demonstrate this, but those developed by Johann Andreas Stein and some Parisian makers of the period have vanished entirely, leaving only the enthusiastic claims of their inventors.

10. THE GEIGENWERK

One basic feature of the harpsichord is that after the string has been plucked the sound dies away fairly quickly; several ornamented keyboard techniques were developed to overcome this deficiency, by producing, in effect, a more sustained sound. A completely different technique for sustaining the sound of a keyboard instrument was used by Hans Haiden in the second half of the 16th century in his invention the 'Geigenwerk'. This instrument has only the keyboard and basic shape in common with the harpsichord since its mechanism is that of a bowed instrument. Haiden was not the first to have thought of this idea, and indeed his instrument is only a larger variant of the hurdy-gurdy which had been known since at least the 13th century; in the diary of Leonardo da Vinci (in the Bibliothèque de l'Institut, Paris; MS H) and also in his drawings in the Codico Atlantico (in the Biblioteca Ambrosiana, Milan) there are bowed keyboard instruments: either the strings are pressed against an endless bow, or the endless bow is applied to the fixed strings.

Haiden produced a working example of his *Geigenwerk* by 1575 and an improved version in 1599, for which he received an imperial privilege in 1601. He described this version in two pamphlets, *Commentario de musicali instrumento* (Nuremberg, 1605), and *Musicale instrumentum reformatum* (Nuremberg, n.d., and 1610). His account in the latter was quoted in full by Praetorius (1618), who also provided the only surviving illustration of the instrument, which resembled a rather bulky harpsichord (see fig.52). At various times Haiden used gut or wire strings, with parchment-covered wire strings in the bass; since the gut strings did not stay in tune, however, he changed to using only metal ones. The bowing action was provided by five parchment-covered wheels against which the individual strings (one for each note) could be drawn by means of the keyboard. The wheels were turned by a treadle. Haiden claimed that the instrument was capable of producing all shades of loudness, of sustaining notes indefinitely, and of producing vibrato.

As late as the second decade of the 18th century, there was a *Geigenwerk* in the Medici collection in Florence, and another at Dresden was examined by J. G. Schröter. An instrument made by Raymondo Truchado in Spain in the first half of the 17th century, and apparently based on Haiden's writings, is in the Brussels Museum of Musical Instruments. Several other inventors also modelled bowed keyboard instruments on Haiden's *Geigenwerk*, or produced similar instruments: the *clavecin-vielle* made by Cuisinié in Paris in 1708 was in fact a large hurdy-gurdy (with six strings and on four legs), in which the wheel was turned by the pedal in such a manner that the keyboard with 29 keys could be played by both hands. It was similar to the *Klaviergamba* built by Georg Gleichmann in Ilmenau around 1725. In 1741 Roger Plenius patented the 'lyrichord' (see fig.53, p.198) in which the strings were tensioned (and thereby tuned) by lead weights and vibrated by the friction of various wheels, each rotating at a different speed. Crescendo and decrescendo were possible, and also vibrato. In 1762 le Gay combined a bowed mechanism with a hammer piano, as did Karl Greiner who built a *Bogenhammerklavier* in 1779. All sorts of experiments continued throughout the 18th century and well beyond: in Paris in 1742 le Voir sought to improve the quality of sound emanating from his bowed keyboard instrument by fitting a cello and a viola into the body of the harpsichord. Efforts in devising such bowed instruments

52. Geigenwerk: woodcut from Praetorius's 'Syntagma musicum' (2/1619)

might be said to have reached a climax with the construction by the Berlin mechanic Johann Hohlfeld in 1754 of a *Bogenklavier*, for which C. P. E. Bach composed a sonata, but instrument makers and inventors continued to develop such instruments until the middle of the 19th century. Even in the 20th, bowed keyboard instruments have been invented, such as Luigi Russolo's *piano enarmonico*, in which the keys brought a moving belt into contact with coiled springs to produce a rich tone.

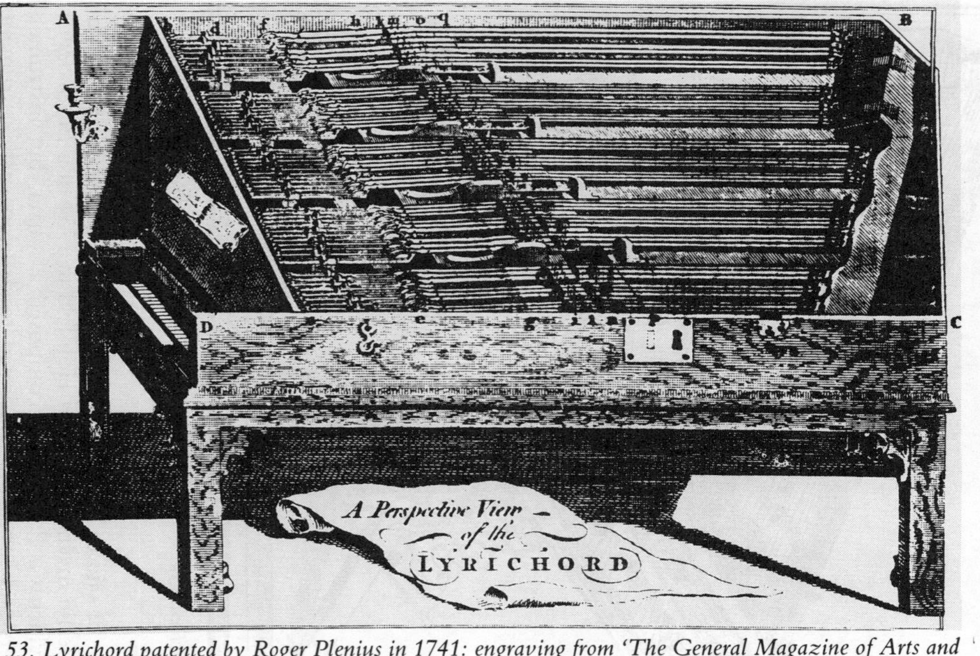

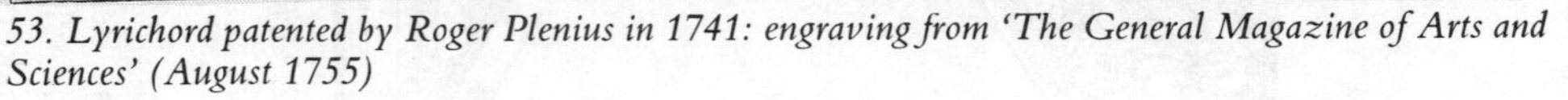
53. *Lyrichord patented by Roger Plenius in 1741: engraving from 'The General Magazine of Arts and Sciences' (August 1755)*

CHAPTER FIVE

Repertory

Before the mid-17th century composers made little stylistic distinction between one keyboard instrument and another, and players used whichever happened to be available or was best suited to the occasion. Liturgically based works and those containing long-sustained notes would be heard most often on the organ, and dances and settings of popular tunes on one or other of the string keyboard instruments; nevertheless, much of the repertory could be shared, specially those organ works that were written for manual(s) only. Because of this, the earlier sections below refer to a number of works that were doubtless intended primarily for organ, but would also have been played on the virginal, harpsichord, spinet or clavichord. They may be described comprehensively as being 'for keyboard'.

With the advance of the 17th century music for the organ became increasingly idiomatic and thus distinct from that intended for the early string keyboard instruments. Furthermore, during the following century the latter were gradually superseded by the more powerful and 'expressive' piano, which had been invented about 1700. By the mid-19th century the early instruments were virtually extinct, and were not revived until interest in them and their music began to reawaken towards the beginning of the 20th century.

1. 14TH AND 15TH CENTURIES

Although the surviving sources of keyboard music go back no further than the early 14th century, players and instruments are known to have existed long before. It therefore seems likely that the lack of an earlier repertory is due partly to the wholesale loss of the manuscripts concerned, and partly to the fact that players

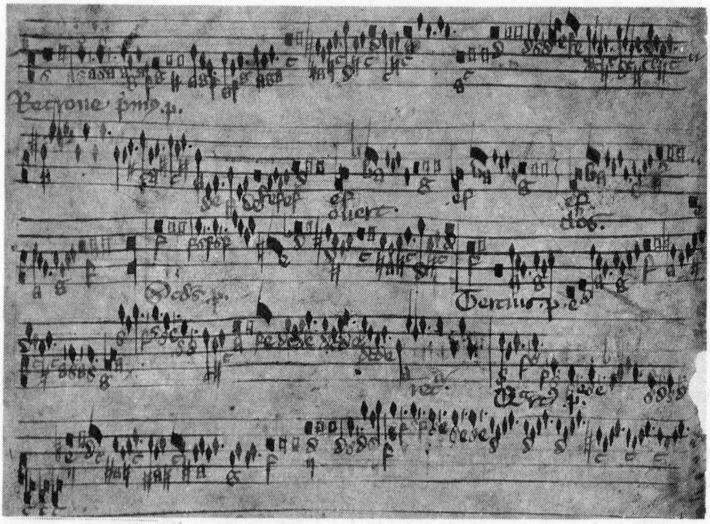

54. *The earliest known example of keyboard tablature: the Robertsbridge Codex, c1320 (f.43v, detail)*

during the earliest period relied largely on vocal originals and improvisation.

The earliest known keyboard source by almost a century is the Robertsbridge Codex of about 1320 (British Library, London, Add.28550). This incomplete two-leaf manuscript from the former priory of Robertsbridge, Sussex, is a curious hybrid, for though it may have been copied in England, it is written in a form of Old German keyboard tablature, and the music it contains is probably either French or Italian in origin. It consists of two and a half dances in the form of *estampies* (melodies without text) and two and a half arrangements of vocal motets, two of which are found in the 14th-century *Roman de Fauvel*. Thus the two main categories of early keyboard music are already represented: namely, purely instrumental works, and works that are derived in some way from a vocal original. In the *estampies* the writing is mostly in two parts, though at cadences the texture tends to become fuller, as often happens in keyboard music. In the motet arrangements the top part of the three-voice original is decorated, or 'coloured', mainly in conjunct motion and in relatively short note values. The remaining parts are generally left unchanged, though occasionally one is omitted or an extra part added. There is no indication of the instrument for which the pieces were intended.

The bulk of the Reina Manuscript (Bibliothèque Nationale, Paris, nouv.acq.fr.6771) and the musical sections of the Faenza Manuscript (Biblioteca Comunale, Faenza, MS 117) belong respectively to the late 14th century and the early 15th. Only a keyboard setting of Francesco Landini's ballata *Questa fanciulla* and an unidentified keyboard piece are included among Reina's otherwise exclusively vocal repertory; but the oldest part of Faenza consists entirely of keyboard pieces, though at one time it was thought they might have been intended for two non-keyboard instruments. There are arrangements of secular vocal works by Italian and French composers of the 14th and early 15th centuries (such as Landini, Jacopo de Bologna, Machaut and Pierre de Molins) and settings of liturgical chants including two Kyrie–Gloria pairs based on the plainsong Mass IV, *Cunctipotens genitor Deus* (see ex.1, the conclusion of a Kyrie verse). The Kyrie–Gloria settings are the first of countless

Ex.1 Faenza MS: Conclusion of a Kyrie verse (. . . . eleison)

plainsong settings designed for *alternatim* performance during the liturgy, in which only the alternate verses are set for organ, while the remainder are sung in unison by the unaccompanied choir. Except for a few three-part cadential chords in Faenza, the pieces in both manuscripts are all in two parts, though many of the secular vocal originals are in three.

The remaining 15th-century sources are all German, three of the most significant being Adam Ileborgh's tablature of 1448 (formerly in the Curtis Institute of Music, Philadelphia; now privately owned), Conrad Paumann's *Fundamentum organisandi* of 1452 (Deutsche Staatsbibliothek, Berlin, Mus.ms.40613), and the Buxheim Organbook of about 1460-70. Ileborgh's tablature is notable for its five short preludes which are the earliest known keyboard pieces (other than dances) that do not rely in any way on a vocal original. In one of them pedals are indicated; and a double pedal part seems to be required in two others, where a florid upper line crosses a pair of lower lines as they move slowly from a 5th to a 3rd and back again. Paumann's *Fundamentum* is one of several treatises that illustrate techniques used in extemporization and composition. It provides examples of a florid part added above various patterns of bass; of decorated clausulas; of two free parts; and of two parts above a static bass. In addition, it includes a number of preludes, and of two- and three-part pieces based on both sacred and secular tenors, by Georg de Putenheim, Guillaume Legrant, Paumgartner and (presumably) Paumann himself. The Buxheim Organbook, which may also be associated with Paumann or his disciples, is the most comprehensive of all 15th-century keyboard sources. It contains over 250 pieces, of which more than half are based on either chansons or motets by German, French, Italian and English composers. They are of two main types. In the first, the whole of the original texture is used, one part being embellished while the rest are left more or less untouched, as in the Robertsbridge motets. In the second, the tenor alone is borrowed, to provide the foundation for what is otherwise a new composition. The rest of the manuscript includes liturgical plainsong pieces, preludes, and pieces based on basse danse melodies. In the liturgical pieces the plainsong generally appears in long equal notes in one part, while the remaining parts have counterpoints in more varied rhythms. But occasionally the plainsong itself is ornamented or even paraphrased. The preludes are mostly regularly barred (unlike Ileborgh's), and often alternate chordal and florid passages in a way that foreshadows the later toccata. Most of the pieces are in three parts, although sometimes in two and occasionally in four (an innovation for keyboard music). The tenor and countertenor lines – the two lowest in the three-part pieces – have roughly the same compass; and as the countertenor was always added last, as in earlier vocal music, it constantly and often awkwardly has to cross and recross the tenor in order to find a vacant space for itself. Pedals are sometimes indicated by the sign *P* or *Pe*; apparently they could also be used elsewhere, for a note at the end of the volume explains that they should always play whichever tenor or countertenor note happens to be the lower.

2. 16TH CENTURY

Printed keyboard music began to appear during the 16th century. Liturgical plainsong pieces remained of paramount importance; but they were joined by settings of Lutheran chorales (hymn tunes) and an increasing number of secular works such as dances, settings of popular tunes, variations, preludes and toccatas. Of great significance, too, were the sectional contrapuntal forms of keyboard music derived from 16th-century vocal forms, including the contrapuntal keyboard ricercare as well as the canzona, capriccio and fantasy.

German sources contain dances and arrangements of both sacred and secular vocal music, some being anthologies while others appear to be the work of a single composer. Although most of them are described as being for either 'Orgel', or 'Orgel oder Instrument', they are generally equally well (or even better) suited to harpsichord or spinet. The two earliest are a pair of manuscripts (Öffentliche Bibliothek der Universität, Basle, F.IX.22 and F.IX.58) written by Hans Kotter between 1513 and 1532 for the use of the Swiss humanist Bonifacius Amerbach. In addition to embellished arrangements of vocal works by Paul Hofhaimer, Heinrich Isaac, Josquin Desprez and others, they include preludes and dances, some of which are by Kotter himself. Typical of the latter is a *Spanioler* in which the basse danse melody *Il re di Spagna* is given to the tenor, each note being played twice in long–short rhythm, while treble and bass have more lively counterpoints. Later tablatures, some printed and others manuscript, are those of Elias Nikolaus Ammerbach (1571, 1583), Bernhard Schmid the elder (1577), Jacob Paix (1583), Christoph Loeffelholz von Colberg (1585) and August Nörmiger (1598). A new trend is shown by the inclusion of 20

Lutheran chorales in Ammerbach's volume and over 70 in Nörmiger's. The plain melody is generally, though not invariably, given to the top part, while the remaining three parts provide simple harmony with an occasional suggestion of flowing counterpoint. A *Fundamentum* of about 1520 by Hans Buchner, similar to Paumann's but dealing with a later style of three-part counterpoint, contains the earliest known example of keyboard fingering.

The dances in the tablatures and other sources are often grouped in slow–quick pairs, such as a passamezzo and saltarello, or a pavan and galliard, in which the second dance (in triple time) may or may not be a variation of the first (in duple). Not infrequently they are based on one or other of the standard harmonic patterns known throughout western Europe, of which the *passamezzo antico* and the *passamezzo moderno* or quadran were the most common.

In Italy the printing of keyboard music began in 1517 with a book of anonymous arrangements entitled *Frottole intabulate da sonare organi*. The mainly homophonic textures of the four-part vocal originals (mostly by Bartolomeo Tromboncino) are lightly embellished to give a more flowing effect; but, as is characteristic of keyboard music, the number of parts employed at any moment depends more on the capacity of a player's hands, and the demands of colour and accent, than on the rules of strict part-writing. Similar freedom was exercised, as illustrated in ex.2, by Marco Antonio Cavazzoni, whose *Recerchari, motetti, canzoni* (1523) was the earliest keyboard publication by a named Italian composer. His brilliant son Girolamo Cavazzoni, perhaps working under the influence of the Spaniard Antonio de Cabezón (see below), developed from his father's rambling ricercares a clearly defined form in dovetailed imitative sections that became the standard pattern of such works. His two books of *intavolature* (1542) contain hymn and plainsong settings for organ and two canzonas with French titles. At least one of the latter, the lively *Il est bel et bon*, is virtually an original composition, for it uses no more than the first bar and a half of the chanson by Passereau on which it is allegedly based.

During the second half of the century the most important centre for Italian keyboard music was Venice, where Andrea Gabrieli, his nephew Giovanni Gabrieli and Claudio Merulo were numbered among the organists of St Mark's Cathedral. Andrea's keyboard works were issued posthumously between

Ex.2 M. A. Cavazzoni: Intabulation of Plus de regres

1593 and 1605 by Giovanni, who added several of his own compositions to his uncle's. Each contributed a set of *intonazioni* in all the 'tones' or modes – short pieces used during the liturgy either as interludes, or to give the choir the pitch and mode of the music they were about to sing. Like earlier preludes, they often include some brilliant passage-work; this led by extension to the toccata, essentially a keyboard piece in several contrasted sections designed to display the varied capabilities of a player and his instrument. The toccatas of Andrea and Giovanni Gabrieli rely mainly on the contrast between sustained writing and brilliant passage-work; but Merulo enlarged the form by introducing one or more sections of imitative counterpoint. In addition to toccatas all three composers wrote ricercares, ornate chanson arrangements and original canzonas. The ricercares follow the sectional pattern established by Girolamo Cavazzoni; but those of Andrea and Giovanni Gabrieli have fewer themes (sometimes only one) and achieve variety by the use of inversion, augmentation, diminution and stretto, and by the importance given to secondary material such as a countersubject or a new thematic tag. Canzonas tend to be lighter in feeling than ricercares, and often begin with a rhythmic formula of three repeated notes, for instance minim–crotchet–crotchet. None of the works requires pedals, and many of them are as well suited to the harpsichord as to the organ.

The earliest Italian keyboard dances are found in a small anonymous manuscript of about 1520 (Biblioteca Nazionale Marciana, Venice, Ital.IV.1227, ed. in *Balli antichi*, 1962). Both here and in the anonymous *Intabolatura nova di varie sorte de balli* (1551), the melody is confined to the right hand, while the left has little more than a rhythmical chordal accompaniment. More sophisticated textures appear in the dance publications of Marco Facoli (1588) and Giovanni Maria Radino (1592), proving that the addition of simple counterpoint and right-hand embellishments can make such pieces sufficiently interesting to be played and heard for their own sake, and not merely as an accompaniment for dancing.

Although England lagged far behind the Continent in printing keyboard music, British composers led the way in developing keyboard techniques. The broken-chord basses characteristic of later string keyboard writing appear in a manuscript of about 1520–40 (British Library, London, Roy.App.58), which contains an adventurous 'Hornepype' by Hugh Aston and two anonymous pieces, *My Lady Careys Dompe* and *The Short Mesure of My Lady Wynkfylds Rownde*, which may also be by him. All three have ostinato left-hand parts.

The principal contributors to the keyboard anthology known as the Mulliner Book (*c*1550–75; British Library, London, Add. 30593) were Thomas Tallis and William Blitheman. In addition to many plainsong pieces the manuscript contains simple transcriptions of Latin and English motets, secular partsongs and consort music. Most of the music was probably intended primarily, though not exclusively, for organ; but three anonymous pieces at the beginning of the manuscript, and a later pavan by Newman (no.116), have the chordal basses that distinguish string keyboard music. Similar basses are found in the Dublin Virginal Manuscript (*c*1570; Trinity College, Dublin, D.3.30), which consists almost entirely of anonymous dances. These contain a sprinkling of the double- and single-stroke ornaments and many of the varied repeats or 'divisions' that later became ubiquitous features of the virginal style.

The only surviving French sources of the 16th century are seven small books of anonymous pieces published by Pierre Attaingnant of Paris in 1530-31. Three are devoted to chanson arrangements (some of them also known in lute versions); two to *alternatim* plainsong settings of the Mass, *Magnificat*, and *Te Deum*; one to motet arrangements; and one to dances (galliards, pavans, branles and basse danses). All are described as being 'en la tablature des orgues, espinettes et manicordions'; but the dances and chanson arrangements are best suited to string keyboard instruments.

The most oustanding keyboard composer of the first half of the century was Antonio de Cabezón, organist to Charles V and Philip II of Spain. A number of his works (ascribed simply to 'Antonio') were included in Venegas de Henestrosa's anthology, *Libro de cifra nueva*(1557); but the principal source is the volume of Cabezón's own *Obras de música* published posthumously in 1578 by his son Hernando. Although both collections are described as being for 'tecla, arpa y vihuela' (keyboard, harp and lute), they were intended primarily for keyboard – the plainsong settings for organ, the *diferencias* (variations) for harpsichord, and the tientos (ricercares) for either instrument. Cabezón's style is severe, with textures that are generally contrapuntal and always in a definite number of parts. The tientos present a number of themes in succession, each section beginning with strict imitation and culminating in free counterpoint, often in relatively small note values. No ornament signs are used, but a favourite embellishment is a written-out shake with turn. Moreover, it seems likely that contemporary players would have added extempore *redobles* (turns), *quiebros* (shakes, and upper or lower mordents), and *glosas* (diminutions), as recommended in Tomás de Santa María's treatise, *Libro llamado Arte de tañer fantasia* (1565). The *diferencias* are lighter in mood, though still strictly contrapuntal. In one of the finest, *El canto llano del caballero*, the melody is at first plainly harmonized, then given successively to soprano, tenor, alto, and again tenor, with flowing counterpoint in the remaining voices. As a member of Philip's private chapel, Cabezón visited Italy, Germany and the Netherlands in 1548–51, and the Netherlands and England in 1554–6; yet he appears to have had surprisingly little influence on the many composers he must have met in his travels.

Keyboard music from Poland survives in several manuscripts, of which the most comprehensive is the so-called Lublin Tablature, copied by Jan z Lublina during the years 1537–48 (Biblioteka Polskiej Akademii Nauk, Kraków, MS 1716). It contains some 250 works, mostly anonymous, and includes liturgical plainsong pieces, preludes, dances (often in slow–quick pairs), and arrangements of vocal works with Latin, German, French, Italian and Polish titles. The influence of the

German school is apparent throughout and extends even to the notation used.

3. 17TH CENTURY

Among the principal forms and types of keyboard music introduced during the 17th century were suites, genre or character-pieces, paired preludes and fugues, chorale preludes, and (from about 1680) sonatas. Superb organs in northern and central Germany encouraged the use of the newly independent pedal registers, thus underlining the difference between organ and string keyboard idioms. But the earlier more 'generalized' style of keyboard writing tended to persist wherever organs were less highly developed.

During the early part of the century the main advances in technique still took place in England, where the printing of keyboard music began at long last with *Parthenia or The Maydenhead of the first musicke that ever was printed for the Virginalls* (1612–13). Its three contributors, Byrd, Bull and Orlando Gibbons, represented successive generations of the great school of virginalists that spanned the late 16th and early 17th centuries. The remaining sources of solo virginal music are manuscripts, however, for the apparent sequel, *Parthenia In-violata* (*c*1624), is for virginal and bass viol. The most comprehensive manuscript source is the Fitzwilliam Virginal Book (*c*1609–19), which provides a cross-section of the whole repertory from Tallis (*c*1505–1585) to Tomkins (1572–1656). Besides containing many unique texts, this remarkable anthology shows the ever-growing popularity of secular works such as dances, settings of song-tunes, variations, fantasias and genre pieces.

Typical of the virginal idiom, as developed by Byrd, are textures that range from contrapuntal imitation to plain harmony in either broken or block chords; a constantly varying number of parts; short figurative motifs; and florid decoration – particularly in the 'divisions', or varied repeats, that are often included in the text. Profuse ornamentation is a constant feature of the style, though oddly enough there is no contemporary explanation of the two signs commonly used to designate ornaments – the double and single stroke.

Keyboard techniques were enormously extended by Bull,

55. Title-page of 'Parthenia' (1612–13)

who was the greatest virtuoso of the day, and by Farnaby, a minor master of rare charm. Brilliant effects were achieved by figuration based on broken octaves, 6ths, 3rds and common chords, by the use of quick repeated notes and wide leaps, and even (in Bull's 'Walsingham' variations, MB, xix, no.85; see ex.3) by the crossing of hands. Farnaby's tiny piece 'For Two Virginals' (MB, xxiv, no.25), one of the earliest works of its kind, consists of no more than a plain and a decorated version of the same music played simultaneously. A clearer grasp of the true principles of duet writing is shown, however, in Tomkins's single-keyboard 'Fancy: for Two to Play' (MB, v, no.32); for though based on choral procedures, its mixture of antiphonal and contrapuntal textures neatly displays the essential individuality-cum-unity of two performers.

Ex.3 Bull: 'Walsingham' variations

By the time the aged Tomkins died in 1656 younger composers were already turning towards a new style, French-influenced, in which the main thematic interest lay in the top line. The change can be seen clearly in the short, tuneful pieces of *Musicks Hand-maide* (1663), a collection of 'new and pleasant lessons for the virginals or harpsycon'. One of the few composers named in it is Matthew Locke, whose more ambitious anthology, *Melothesia* (1673), is prefaced significantly by 'certain rules for playing upon a continued-bass'. It includes seven of his own pieces (voluntaries) for organ and 'for double [i.e. two-manual] organ', and a number of suites (not so named) by himself and others, consisting generally of an almain, corant, saraband and one or more additional movements. Similar suites were written later by Blow and his pupil Purcell, the principal contributors to *The Second Part of Musick's Hand-Maid* (1689); Purcell's were issued posthumously as *A Choice Collection of Lessons for the Harpsichord and Spinnet* (1696) and four of Blow's appeared two years later with the same title. All these publications were aimed at the amateur. But Purcell's harpsichord music, though small in scale, is no less masterly than his more ambitious works for theatre, court and the church; and at times it achieves a depth and poignancy – particularly in the ground basses of which he was so fond – that is quite disproportionate to its size.

Musical exchanges between the Continent and England had taken place during the early years of the century. Arrangements of madrigals by Marenzio and Lassus and original works by Sweelinck, organist of the Oude Kerk in Amsterdam, were included in the Fitzwilliam Virginal Book; and even more significantly, Bull, Peter Philips and other Catholic recusants found refuge in the Netherlands and elsewhere, and thus spread abroad the advanced English keyboard techniques. Sweelinck himself was much influenced by the innovations, as can be seen not only from his harpsichord works, but also from his organ variations on Lutheran chorales and his echo-fantasias. Although none of his keyboard works appeared in print, Sweelinck's fame as the foremost teacher in northern Europe brought him numerous pupils, particularly from the neighbouring parts of Germany. The latest techniques were thus passed on to a younger generation of composers, who in their turn carried them still farther afield.

German composers of the period may conveniently be divided into two groups: those who worked in the Protestant north and centre; and those of the Catholic south, including Austria. To the former group belong Sweelinck's pupils, Scheidt and Scheidemann. Scheidt's keyboard works were issued in two collections, the *Tabulatura nova* (1624) and the *Tabulatur-Buch hundert geistlicher Lieder und Psalmen* (1650). (In the first of these the description 'new' refers to the use of open score in place of letter notation.) The later volume consists of simple four-part settings of Lutheran chorales for accompanying unison singing. One of the sets of variations for harpsichord is based on the English song *Fortune my Foe*, which was also set by Sweelinck, Byrd and Tomkins. Scheidemann's works, like those of most northerners, remained unpublished. The most outstanding of all the northerners was, however, Buxtehude, who left his native

Denmark in 1668 to become organist of the Marienkirche in Lübeck. Although he was primarily an organ composer, the publication in 1941 of the Ryge Manuscript (Det Kongelige Bibliotek, Copenhagen, C.11.49.4°) made available his suites and variations for clavichord or harpsichord; these are so similar in style to those of Nicolas-Antoine Lebègue that the editor did not notice the inclusion of one of Lebègue's suites in the Buxtehude manuscript.

The earliest and most significant southern German composer was Froberger, who, though born in Stuttgart, held the post of court organist in Vienna for 20 years. His ricercares, canzonas and fantasias are strongly influenced by his master, Frescobaldi, but his toccatas are less Italian in style. Although they begin with the usual sustained chords and brilliant flourishes (see ex.4), they generally include two fugal sections on rhythmic variants of a single subject, each section being rounded off with further flourishes. His suites are in an expressive, romantic vein better

Ex.4 Froberger: Toccata no. 1

suited to the clavichord than to the harpsichord. They are French in style, and are said to have been the first to establish the basic suite pattern of four contrasted national dances: i.e. an allemande (German), courante (French) or corrente (Italian), sarabande (Spanish) and gigue or jig (English). In Froberger's autographs the gigue either precedes the saraband or is omitted altogether; but when the works were published posthumously (Amsterdam, 1693) the order was changed ('mis en meilleur ordre') and the gigue placed at the end. During the last ten years of his life Froberger travelled widely in Germany, France, the Netherlands and England, meeting Chambonnières and Louis Couperin in Paris and Christopher Gibbons (son of Orlando) in London; thus he too played a significant part in the cross-fertilization of national styles.

Among the lesser southerners were Alessandro Poglietti, Georg Muffat and J. C. F. Fischer. Although Poglietti was probably an Italian, he became court organist in Vienna shortly after Froberger, and in 1677 presented Leopold I and his empress with an autograph collection of his harpsichord pieces entitled *Rossignolo*. Besides a ricercare, a capriccio and an *Aria bizarra*, all based on the *Rossignolo* theme, it includes a virtuoso 'imitation of the same bird', and an *Aria allemagna* with 20 variations. Each of the latter has an illustrative title ('Bohemian Bagpipes', 'Dutch Flute', 'Old Woman's Funeral', 'Hungarian Fiddles' etc), and in number they match the age of the empress, to whom they were dedicated. Muffat's *Apparatus musico-organisticus* (1690) includes four harpsichord pieces of which the large-scale Passacaglia in G minor and the shorter Ciacona in G have a power and breadth more typical of the north than of the south. In contrast to these, the four collections by Fischer are wholly southern in their delicacy of feeling. *Les pièces de clavessin* (1696) and the *Musicalischer Parnassus* (1738) are devoted to harpsichord suites, each of which begins with a prelude of some sort and continues with a group of dances or other pieces, not always including the usual allemande, courante, sarabande and gigue. The other two volumes, *Ariadne musica* (1702) and *Blumen Strauss* (1733), contain miniature preludes and fugues. The *Ariadne* group interestingly foreshadows Bach's *Das wohltemperirte Clavier* in the wide range of its key scheme, and even in some of its themes (Fischer's eighth fugue in E obviously inspired Bach's ninth from book 2).

In Italy the main centre for keyboard music moved from Venice to Naples and then to Rome. From Ascanio Mayone's *Diversi capricci* (1603 and 1609) and Giovanni Maria Trabaci's *Ricercare* (1603 and 1615) it can be seen that although the Neapolitans retained the strict contrapuntal style of the Gabrielis in their ricercares, they broke new ground in toccatas by shortening the sections, increasing their number and heightening the contrast between one section and the next. The same distinction was made by Frescobaldi, who, as organist of St Peter's in Rome, was the most widely acclaimed player and keyboard composer of the day. Although he visited the Netherlands in 1607, when the 45-year-old Sweelinck was at the height of his powers, he was little influenced by the techniques of the north. His works were published during the next 35 years in a series of ten volumes of which some are revised and enlarged editions of others. The three definitive collections are *Il primo libro di capricci, canzon francese e recercari* (1626) and the *Toccate d'intavolatura di cimbalo et organo* with its sequel *Il secondo libro di toccate* (both 1637). (The first two contain important prefaces by the composer concerning interpretation.) Most of the toccatas, capriccios and canzonas in these collections are equally suited to harpsichord and organ, for though some have a primitive pedal part, it generally consists of no more than long-held notes that are already present in the left hand. The works intended primarily for harpsichord include dances (sometimes grouped in threes, with the opening balletto serving as theme for the following corrente and passacaglia), and sets of variations or partitas, a number of which are based on harmonic patterns such as the romanesca and the Ruggiero. The ricercares and plainsong pieces are essentially organ music, as is the liturgical *Fiori musicali* (1635), of which Bach possessed a manuscript copy.

One of the few 17th-century Italian publications devoted wholly to dances was Giovanni Picchi's *Intavolatura di balli d'arpicordo* (1621). Besides the customary passamezzo, saltarello and padoana (pavan), it includes imitations of alien idioms such as a 'Ballo alla polacha' a 'Ballo ongaro' and a 'Todesca'. The corantos in Michelangelo Rossi's *Toccate e correnti d'intavolatura d'organo e cimbalo* (*c*1640) are in a lighter, more tuneful style, though his toccatas are still closely related to Frescobaldi's. This new style can be seen even more clearly in the works of Bernardo Pasquini, who was among the first to apply the title 'sonata' to solo keyboard music. Originally it denoted no more than a 'sound piece' as opposed to a 'sung piece' or 'cantata', for it was applied indiscriminately to toccatas, fugues, airs, dances and suites. But Pasquini, following the example of Corelli's ensemble sonatas, also gave the title to solos in more than a single movement. Among his other works are 15 sonatas for two harpsichords, in which each part consists rather oddly of no more than a figured bass (British Library, London, Add.31501). The 40-odd toccatas of Alessandro Scarlatti are of interest mainly because each contains at least one *moto perpetuo* section, thus anticipating the much later *moto perpetuo* type of toccata.

Much French keyboard music of the 17th century appeared in

print while the composers were still alive; and as the title-pages generally specified either organ or harpsichord, but not both, there is rarely any doubt about the instrument intended. The mid-century saw the emergence of the distinctive French harpsichord idiom that exercised a potent influence throughout Europe. In essence it was based on the richly ornamented and arpeggiated textures of lute music. The founder of the school was Chambonnières, who late in life published two books of *Pièces de clavessin* (1670) containing 60 dances grouped according to key. The commonest types are allemandes, courantes (often in sets of three) and sarabandes; occasionally a gigue or some other dance is added. More of his pieces survive in the Bauyn Manuscript (Bibliothèque Nationale, Paris, Res.Vm^{7}.674–5), which also contains almost all the compositions of his pupil Louis Couperin, the one outstanding French keyboard composer who never saw any of his own works in print. In addition to the forms used by his master, Couperin wrote a number of 'unmeasured preludes' of a type peculiar to France. Another pupil of Chambonnières was Jean-Henri D'Anglebert, whose *Pièces de clavecin* were published in 1689. The volume is unusual in two respects, for it includes five fugues for organ, and 15 of its 60 harpsichord pieces are arrangements of movements from operas by Lully. D'Anglebert's magnificent *Tombeau de Mr. de Chambonnières* is a good example for keyboard of a type of memorial composition of which French composers have always been specially fond.

4. 18TH CENTURY

All the forms employed during the 17th century remained in use during the first half of the 18th; but sonatas (of other than the classical type) acquired increasing importance, and ritornello form (derived from the Neapolitan operatic aria) provided the foundation on which every concerto and many extended solo movements were built.

French keyboard composers were untouched by these developments, however, and continued to confine themselves to dances and genre pieces for harpsichord, and to short liturgical and secular works for the organ. The two outstanding figures among them were Louis Couperin's nephew François Couperin

PIECES

DE

CLAVECIN

COMPOSÉES

PAR

Monsieur Couperin

Organiste de la Chapelle du Roy, &c.

Et Gravées par du Plessy.

PREMIER LIVRE.

Prix 10.tt en blanc.

A PARIS

Chés L'Auteur.
Chés Le Sieur Foucaut rue St. honoré à la Regle d'or.

1713.

Avec Privilege de sa Majesté

Gravé par Berey

56. Title-page of François Couperin's 'Pieces de clavecin . . . premier livre' (Paris, 1713) the younger and Jean-Philippe Rameau, a near-contemporary of Bach. François Couperin's four books of *Pièces de clavecin* (1713–30) are the crowning achievement of the French clavecin school. The 220 pieces range from elegant trifles to the majestic Passacaille in B minor (*ordre* no.8) and the sombre allemande *La ténébreuse* (*ordre* no.3), which is almost too intense in mood for the dance form in which it is embodied. Couperin's views on teaching, interpretation, ornamentation and fingering are set forth in his *L'art de toucher le clavecin* (1716, rev. 2/1717), a fascinating treatise which nevertheless often fails to answer questions that remain puzzling. Rameau's instructions to the player are contained in two of the prefaces to his four books of harpsichord pieces issued between 1706 and 1741. The works are generally simpler in texture and less richly ornamented than Couperin's, but more adventurous harmonically and in their use of the keyboard. The composer himself noted that it would take time and application to appreciate the (harmonic) beauty of parts of the piece entitled *L'enharmonique*; and he provided fingering for the widely spaced left-hand figure in *Les cyclopes* because of its unusual difficulty. Rameau's final keyboard publication, *Pièces de clavecin en concerts* (1741), is primarily a collection of five suites for violin, bass viol and harpsichord, but it also includes a solo harpsichord version of four of the movements. This practical plan was anticipated, though in reverse, in Gaspard Le Roux's *Pièces de clavecin* (1705). There the main works are suites for harpsichord solo, while the arrangements consist of selected movements for trio (instruments unspecified), and several for two harpsichords, the latter being the earliest known French works for that medium. Composers other than Couperin who wrote for both harpsichord and organ include Marchand, Clérambault, Dandrieu, Dagincour and Daquin. Most of their works are in the customary forms; but volumes by Dandrieu (1715) and Daquin (1757) are devoted to sets of variations on popular Christmas melodies, entitled noëls, a type which first appeared in Lebègue's *Troisième livre d'orgue* (*c*1685).

One of the greatest of all harpsichord composers was the Italian Domenico Scarlatti, son of Alessandro and exact contemporary of Bach and Handel. The last 35 years of his life were spent in the service of Maria Barbara of Braganza, at first in Portugal and later in Spain; during that period he appears to have written almost all his 555 single-movement sonatas. Apart from a volume of 30 *Essercizi per gravicembalo* (1738), published under his own supervision, the main sources of his works are two contemporary manuscript collections (Biblioteca Nazionale Marciana, Venice, It.iv.199–213; and Conservatorio di Musica Arrigo Boito, Parma, AG 31406–20), the first of which was copied for his royal patron. Their contents are similar but not identical, and it has been suggested by Ralph Kirkpatrick (*Domenico Scarlatti*, Princeton, 1953) that the order of their contents is to a large extent chronological, and that more than two-thirds of the sonatas were, as the manuscripts indicate, originally grouped in pairs, or sometimes in threes, according to key (this order is retained in Kirkpatrick's facsimile edition, New York and London, 1972, and in Kenneth Gilbert's excellent complete edition, Paris, 1971–84). Although Scarlatti rarely used any structure other than binary form, and seldom aimed at emotional extremes, he achieved an astonishing variety within those self-imposed limits. Moreover he exploited the keyboard in ways never imagined by any of his contemporaries. In the later works he virtually abandoned his wilder flights of hand-crossing; but he never lost his command of both sparkling brilliance and an unexpected vein of reflective melancholy, his delight in technical and harmonic experiment, and his love for the sounds and rhythms of the popular music of Spain.

Scarlatti's followers in Portugal and Spain, among whom were Seixas and Soler, wrote numerous single-movement sonatas similar in style to his own; but as an expatriate he exercised little influence on Italian composers, whose sonatas are of several different types. Those by Della Ciaia (1727) are not unlike sectional toccatas; Durante's (*c*1732) each contain a *studio* in imitative counterpoint followed by a brilliant *divertimento;* Marcello's (MS) are in either three or four movements; and Zipoli's (1716) include liturgical and secular pieces for organ as well as suites and variations for harpsichord. Also intended for either instrument are G. B. Martini's two volumes of sonatas (1742, 1747), the first devoted to two- and three-movement works, and the second to five-movement works that combine features of both the *sonata da camera* and the *sonata da chiesa*.

English keyboard composers during the post-Purcell period rarely rose above a level of honest competence. Tuneful airs and lessons, sometimes grouped into suites, appeared in serial anthologies such as *The Harpsichord Master* (1697–1734) and *The Ladys Banquet* (1704–35), among whose contributors were Jeremiah Clarke, William Croft and Maurice Greene. In addition, separate volumes were devoted to works by Philip Hart, Clarke, Thomas Roseingrave and Greene. Although Croft was not accorded that distinction, he was the most accomplished composer of the group and the only one to come within hailing distance of Purcell. Indeed, the Ground from his Suite no.3 in C minor is actually ascribed to Purcell in one source.

A Scarlatti cult was at one time fostered in England, first by Roseingrave's edition of *XLII suites de pièces pour le clavecin* (1739), which added 12 more Scarlatti sonatas to the 30 published a year earlier in the *Essercizi*; and secondly by Charles Avison's arrangement of a number of the sonatas as Twelve Concertos (1744) for strings and continuo.

Of far greater significance to English musical life, however, was the arrival of Handel, who settled in London in 1712 after a successful visit two years earlier. Although at first occupied mainly with Italian opera and later with oratorio, he was obliged to publish his [8] *Suites de pièces de clavecin* (1720) in order to counteract the many 'surrepticious and incorrect copies' that were circulating in manuscript. Other collections of his pieces, all unauthorized, appeared later in London and Amsterdam. Some of the suites follow the normal pattern of allemande–courante–sarabande–gigue; but more often they include Italianate allegros, andantes etc, or consist of nothing else. His keyboard works combine relaxed informality with masterly rhetoric in a way that doubtless reflects the improvisations for which he was famous.

Meanwhile in Germany the way had been prepared for the greatest of all pre-classical keyboard composers, J. S. Bach. Among his many musical ancestors, other than relatives, the most significant was Buxtehude (see above). So great was Bach's reverence for him that in 1705 he walked from Arnstadt to Lübeck in order to hear his Abendmusiken – the yearly choral and instrumental performances given on the five Sundays before Christmas. Somewhat less influential were Kuhnau and Georg Böhm. The keyboard works of Kuhnau, Bach's predecessor at the Thomaskirche in Leipzig, include two notable volumes: firstly, the *Frische Clavier Früchte, oder sieben Suonaten* (1696), the earliest publication in which the title 'sonata' is given to a solo as distinct from an ensemble work; and secondly, [6] *Musicalische Vorstellungen einiger biblischen Historien* (1700), the 'musical representations of biblical stories' that provided the model for Bach's early *Capriccio sopra la lontananza del suo fratello*

XLII
Suites de Pieces
Pour le
CLAVECIN.
En deux Volumes.
Composées par
Domenico Scarlatti
Vol: I

NB. I think the following Pieces for their Delicacy of Stile, and Masterly Composition, worthy the Attention of the Curious, Which I have Carefully revised & corrected from the Errors of the Press.
Thos. Roseingrave.

LONDON
Printed for, and sold by B: Cooke at the Golden Harp in New street Covt. Garden; Where may be had Volume the 2d.
this work contains 14 pieces more than any other Edition hitherto extant. 12 of which, are by this Author, ye other 2 is over & above ye No. propos'd

57. Title-page of Domenico Scarlatti's 'XLII Suites de pièces pour le clavecin', i (London: Cooke, 1739), edited by Roseingrave

dilettissimo BWV992. The influence of Böhm, though conjectural, would have been earlier and more direct, for he was organist of the Johanniskirche in Lüneburg when Bach was a choirboy at the nearby Michaeliskirche. Böhm's sensitive suites in the French style for clavichord or harpsichord were unpublished, but the evidence of Bach's own works suggests that he must have been familiar with them as a boy.

A near-contemporary of Bach and Handel, and a friend of both, was the prolific Telemann. The admiration of the two slightly younger men for his music can best be understood by reference to works such as the *XX kleine Fugen* (1731).

Comparatively few of Bach's own keyboard works were published during his lifetime. The most comprehensive collection, the *Clavier-Übung*, was issued in four parts between 1731 and 1742, of which the first, second and fourth contain compositions for both single- and double-manual harpsichord, while the third is mainly devoted to the organ.

The great Passacaglia and Fugue in C minor BWV582 and the six trio sonatas BWV525–30, which were originally written as instructional works for Bach's eldest son, Wilhelm Friedemann, are described merely as being 'for two manuals and pedals', so it remains uncertain whether they were intended primarily for organ or for a harpsichord fitted with pedal-board (such as was used by organists for home practice).

Much of Bach's music for normal harpsichord and/or clavichord was also didactic in aim. The 15 two-part inventions and 15 three-part sinfonias BWV772–801 were first included in a manuscript collection of keyboard pieces for Wilhelm Friedemann dated 1720, and were described in a revision of 1723 as showing not only how 'to play clearly in two voices but also, after further progress, to deal correctly and well with three obbligato parts . . . and above all to achieve a singing style in playing'. Friedemann's book also contained early versions of 11 of the preludes from the first book of *Das wohltemperirte Clavier* (1722), a more advanced collection of 24 preludes and fugues in all the major and minor keys 'for the use and profit of young musicians desiring to learn, as well as for the pastime of those already skilled in the study'. The second book, containing a further 24 preludes and fugues, was not completed until 1744. Two other manuscripts, dated respectively 1722 and 1725, were compiled for the use of Bach's second wife, Anna Magdalena. The first contains five of the six French suites BWV 812–17, each consisting of the usual allemande, courante, sarabande and gigue, with one or more additional dances (*Galanterien*) following the sarabande. The six so-called 'English' suites BWV 806–11 and six partitas BWV825–30 are on a larger scale, for each begins with a prelude of some sort. Those of the English suites (with the exception of no.1) are ritornello-type movements, while those of the partitas are in various forms. The partitas were published singly between 1726 and 1730, and complete in 1731 as part 1 of the *Clavier-Übung*, of which part 2 (1735) consists of the Italian Concerto BWV971 and the French Overture BWV831 (sometimes known as the Partita in B minor), both for two-manual harpsichord. Part 4 (1742), also for two-manual harpsichord, is devoted to a single work: the monumental Aria with 30 Variations BWV988, usually known as the Goldberg Variations, which Tovey described as 'not only thirty miracles of variation-form, but . . . a single miracle of consummate art as a whole composition'.

Slightly later in date is the *Musikalisches Opfer* BWV1079, a collection of fugues, canons etc for various instruments on a theme provided by Frederick the Great. It includes two ricercares for solo keyboard, of which the second, in six parts, was originally printed in open score. This was not an unusual method of presenting keyboard music when its aim was partly didactic. It was used again for Bach's posthumous treatise *Die Kunst der Fuge* BWV 1080, in which the majority of the fugues are clearly intended for solo keyboard, though they have frequently been arranged for various ensembles in the 20th century.

During the Weimar period Bach made solo keyboard versions, some for organ and others for harpsichord, of 22 concertos by various composers, including Vivaldi, Marcello and Telemann. These paved the way for his later concertos for solo harpsichord and strings BWV1052–8, which were the first of their kind (and roughly contemporary with Handel's organ concertos). All seven are arrangements of earlier concertos of his own – mostly for solo violin and strings – several of which have not survived. The only original keyboard work in this form appears to be the Concerto in C for two harpsichords and strings BWV1061; the remaining two for the same medium, and those for three and four harpsichords and strings, are also arrangements of concertos originally by either Bach himself or (in one case) Vivaldi.

In its depth and range of emotion, contrapuntal skill and

perfection of design, Bach's keyboard music far surpasses that of any of his contemporaries or predecessors; yet by the time of his death it was generally regarded as old-fashioned. The Baroque era had ended; the contrapuntal style was outmoded; and the harpsichord and clavichord were beginning to make way for the fortepiano, which combined the power of the one with the sensitivity of the other.

The gradual change can be seen in the works of three of Bach's sons. The eldest, Wilhelm Friedemann, continued to write fugues; but his more characteristic polonaises and three-movement sonatas were typical of the new *empfindsamer stil* (expressive style), which exploited sudden and violent dynamic changes impractical on the harpsichord but readily realizable on either clavichord or fortepiano. His brother Carl Philipp Emanuel was the chief keyboard exponent of the style and exercised a wide influence through his numerous sonatas, fantasias, rondos, etc, and his important treatise *Versuch über die wahre Art das Clavier zu spielen* (1753–62). The youngest brother, Johann Christian (known as 'the London Bach'), was an enthusiastic advocate of the fortepiano, which he used as an orchestral continuo instrument (in place of the harpsichord) after he had settled in England in 1761.

Needless to say, the earlier instruments long remained in use domestically, so it is not surprising that astute publishers included the names of both the old and the new on their title-pages. Thus Johann Christian's Six Sonatas, op.5, first published in London in 1766, were announced as being 'for Piano Forte or Harpsichord'. (Even as late as 1802 the first edition of Beethoven's so-called 'Moonlight' Sonata, op.27 no.2, was improbably described as 'per il Clavicembalo o Piano-Forte'.) Most of Philipp Emanuel's title-pages used the generic word *Clavier* (keyboard), which in Germany included harpsichord, clavichord, fortepiano, and even the manual(s) of an organ. Much of his music is ideally suited to the sensitive clavichord; but it generally requires a five-octave instrument of the type made by Friederici (*F'* to *f'''*), since he often exceeds the more usual four-octave compass (*C* to *c'''*). Philipp Emanuel occasionally used the more precise term 'per il cembalo' (for the harpsichord); but it seems likely that the word was then beginning to acquire the wider meaning of *clavier*, for his *Sei sonate per cembalo*, WQ49 (1744), contain dynamics that cannot be realized on a harpsichord. Similarly, Haydn's Sonata in C minor, Hob.XVI/20

(1771), was described on the autograph by the composer himself as 'per il Clavi Cembalo', and in the first edition (Artaria; Vienna, 1780) as 'per il Clavicembalo o Forte Piano'. No new music for either harpsichord or clavichord appears to have been written during the 19th century.

5. 20TH CENTURY

The interest of 20th-century composers in the harpsichord and clavichord lagged behind the early-instrument revival, which was already well under way by the beginning of the century. Among the earliest solo pieces to appear was Delius's *Dance for Harpsichord* (1919), which, though dedicated to the harpsichordist Mrs Violet Gordon Woodhouse, is quite unplayable on the instrument. (Its many dynamics are impractical and it frequently relies on a non-existent sustaining pedal.) Ensemble works, achieved or projected, include Walter Piston's Sonatina (1945) for violin and harpsichord, and the unwritten Sonata by Debussy for oboe, horn and harpsichord. Among larger-scale works are concertos by de Falla (1923–6), Poulenc (*Concert champêtre*, 1927–8), Walter Leigh (*Concertino*, 1936), and Frank Martin (1952), together with Martin's *Petite symphonie concertante* (1945) for harpsichord, piano, harp and strings, and Elliott Carter's Double Concerto (1961) for harpsichord, piano and orchestra. Some of these works are unplayable on present-day reproductions of 17th-century harpsichords, for they were originally written for the famous player Wanda Landowska, whose powerful Pleyel instrument was equipped with seven pedals to allow instantaneous changes of registration, in place of hand-stops that take an appreciable time to operate.

The 20th-century repertory for clavichord is far less extensive, doubtless because the instrument's limited power unsuits it to both ensemble music and the concert hall. Of the few solo works that exist, specially notable are Herbert Howells's *Lambert's Clavichord*, op.41 (1926–7) and *Howells' Clavichord* (1951–61), both designed for a four-octave instrument. They consist of musical 'portraits' of friends couched in terms of virginalist dance forms, and show true understanding of the clavichord's capabilities.

APPENDIX ONE

Glossary of Terms

Cross-references within this appendix are distinguished by the use of small capitals, with a large capital for the initial letter of the entry referred to, for example:

See KEYBOARD.
Clavier (Fr.). KEYBOARD.

Action. The mechanism by means of which the strings or pipes of a keyboard instrument are sounded when a key is depressed.

Arcade. The decoration applied to the front of a natural key. Usually it is a half-circle cut into the key, but embossed paper or leather may be glued on to the front (e.g. as in many Flemish instruments).

Arpicembalo (It.). The name given to a harpsichord-piano (see Chapter Four, §5) made by Cristofori.

Arpicordo leutato (It.). A gut-strung arpicordo that sounded like a lute (see Chapter Four, §7).

Arpichordum. A device in a harpsichord or virginal in which small metal pins are brought close to the vibrating strings which jar against them, producing a buzzing sound. It is usually found on centre-plucking virginals (muselars), but only for the straight part of the bridge (i.e. from *C/E* to *f′*). According to Praetorius (1618), it imparted 'a harp-like sound'.

Axle. The pin in the tongue of a jack. It holds the tongue in position and acts as a pivot, allowing the tongue to swing forwards and backwards (see fig.1*b*). The axle was commonly a dressmaker's pin, a mass-produced item in the 16th century.

Bach disposition. A term that originates from the disposition of a much-altered harpsichord (no.316 in the collection of the Staatliches Institut für Musikforschung, Berlin) that was thought to have been used by J. S. Bach. The harpsichord has 1 x 16′, 1 x 8′ on the lower manual and 1 x 8′, 1 x 4′ on the upper keyboard; it has been shown however that the 16′ stop is not original, but derives from a 19th-century modification. Many modern German harpsichords have been based on this disposition (see Chapter One, p.99).

Bachklavier. A type of harpsichord developed in the early 20th century by Karl Maendler of Munich. It was intended to be capable of dynamic nuance through touch. Production ceased about 1933 (see Chapter One, p.105).

Backrail. A cloth-padded rail on the keyboard frame on which the ends of the key-levers (furthest from the player) rest (see fig.14).

Balancement (Fr.). *See* BEBUNG.

Balance pin. An iron or brass pin (about 2–4 mm in diameter) which passes through the key and is driven into the balance rail. It serves to define the fulcrum of a key-lever and to guide it in the vertical plane.

Balance point. The point at which a key-lever balances. In many instruments this is not identical with the position of the BALANCE PIN, which is placed so that the key-lever falls back into a horizontal position under its own weight. In order to lighten the touch, the keys on some harpsichords, notably 18th-century French, are balanced so that they fall forward (as if a key were played and held down) when the weight of the jacks is not on them.

Balance rail. A strip of wood in the key bed into which the balance pins are driven. It is usually made of a hardwood, such as beech.

Barring. A general term for the arrangement of ribs or bars used to support and divide off areas of the soundboard.

Baseboard. *See* BOTTOM.

Bebung (Ger., from *beben*: 'to shake', 'to tremble'; Fr. *balancement*). A vibrato effect obtained on the clavichord by alternately increasing and decreasing the pressure of the finger on the key. It is described by a number of mid-18th-century writers, notably F. W. Marpurg and C. P. E. Bach; the latter wrote that it should be delayed until the second half of the note and that it should be used on long *affetuoso* notes.

Belly rail [header]. A piece of wood glued between the baseboard and the front edge (i.e. nearest the player) of the soundboard. It is usually stiffened on the upper (soundboard) edge with another rail; sometimes it is made of two separate pieces of wood, so that the air space under the soundboard is not closed off.

Bentside. The curved side of a harpsichord. It is on the right as seen by the player, between the cheekpiece and the tail.

Bentside spinet ['wing-shaped', 'leg of mutton-shaped' spinet] (Ger. *Querspinett*). A type of spinet in which the strings run at an angle of about 20° from the keyboard to the player's right. One of the case sides is curved in towards the bridge, as in the harpsichord bentside (see fig.35).

Bichord-strung. A stringing in which the strings are paired and the two strings of each pair can usually be sounded only simultaneously.

Most clavichords are bichord-strung. Harpsichords with two 8′ registers are not generally described as bichord-strung since the registers can (usually) be played independently. Early pianos are bichord-strung (often with trichord strings from about *b*′ to the top of the range), but may also have an arrangement that allows one of a pair of strings to be struck; *see* UNA CORDA.

Bottom [baseboard, bottom boards]. In traditional harpsichord construction the underside of the instrument, by which the volume of air under the soundboard is enclosed. It is an important structural member, usually about 12 to 18 mm thick. Some modern designs dispense with a bottom and rely on the frame members for structural strength.

Boxslide. A type of JACKSLIDE. It is in one piece, about 2.5 to 5 cm deep and the guiding slots are continuous for the depth of the slide. It was commonly used in Italian harpsichords; it is also found in bentside spinets and Italian polygonal virginals. (See also Chapter One, p.11.)

Brace. Any form of structural member which has the function of resisting deformation of the case. The term is often applied to frame members that run from the LINER to some other part of the case.

Bray. In German renaissance harps, the right-angled wooden peg that fixed the strings to the pegboard. When plucked the strings vibrate against the brays, producing a buzzing sound. The ARPICHORDUM stop produces a similar effect in a string keyboard instrument.

Bridge. A general term for the member over which the strings run. It usually denotes only the bridge on the soundboard which transmits the vibrations of the strings (*see also* NUT and fig.1). Since the bridge has to hold pins (past which the strings run) it is usually made of hardwood: pear (or similar woods of the *Sorbus* family) and beech were widely used. Italian bridges were sometimes made of cypress. In some early clavichords the several small bridges are not glued to the soundboard but held in place by string pressure.

Bridge pin. A small iron or brass pin (about 0.9–1.3 mm in diameter) which is driven into the bridge. It defines the sounding length of the string and provides a durable point at which the string can bear on the bridge and transmit its vibrations to the soundboard.

Bristle. A stiff hair, usually from the pig, which is used as a SPRING in a jack.

Broken octave. In keyboard instruments, a term used to designate a variation of the SHORT OCTAVE. In the broken octave the keys of the lowest accidentals ('sharps') are divided in order to provide a note that would otherwise be missing. The front part of the divided sharp usually sounds the pitch that would be expected of it in a normal short octave; the back part provides the accidental that would be found in a chromatic octave. Keys divided for this purpose should not be confused with those divided to provide enharmonically-equivalent notes (*see* ENHARMONIC KEYBOARD).

Buff batten [buff stop batten]. The movable strip of wood to which the pads of the BUFF STOP are fixed (see fig.14).

Buff pad. A small pad of soft leather or felt which is held in contact with the string close to the nut.

Buff stop [harp stop] (Fr. *jeu de luth*; Ger. *Lautenzug*; It. *liuto*). A device that mutes the strings by pressing a piece of soft leather or felt against them at the nut, thereby damping out the higher harmonics; the sound takes on a pizzicato quality. It is found on harpsichords of all periods (though rarely on Italian instruments) and on some pianos, especially square pianos of the 18th and early 19th centuries. In harpsichords, the buff stop usually consists of a sliding batten fitted with separate pieces of leather or felt and is usually held against the nut by small pins which are driven into the wrest plank (see fig.14). In 17th-century Flemish single-manual harpsichords the buff batten was often divided into separate treble and bass sections. The buff stop should not be confused with the PEAU DE BUFFLE.

Cap moulding. The moulding on the top edge of a harpsichord case; it is usually found only in Italian harpsichords.

Case. The part of a harpsichord, virginal, clavichord or similar instrument which defines the outside dimensions and shape. It has a major structural role in resisting the tension of the strings. Its acoustical contribution is widely, but erroneously, held to be of importance; it contributes only marginally to the acoustical character, the soundboard being far more important.

Cembal d'amour. A type of clavichord invented by Gottfried Silbermann in the first quarter of the 18th century (see Chapter Three, p.163).

Cembalo (Ger., It.). Harpsichord.

Cembalo traverso (It.). (1) A term for the spinet.

(2) The name used to describe an instrument, made by Cristofori and others, that is in effect a large BENTSIDE SPINET.

Centre plucking. *See* PLUCKING POINT.

Cheekpiece. The short case side to the right of the player as he sits at the instrument.

Chinoiserie. A style of painted decoration in imitation of Chinese or Japan lacquer work; it is usually black, or black and gold. It was a popular form of case decoration in the 17th and 18th centuries (see fig.45).

Choir. A term for a set of strings playable from the keyboard. *See also* REGISTER and STOP.

Chromatic bass octave. In a keyboard instrument, a bottom octave in which no accidentals are missing. The earliest keyboards in harpsichords did not have chromatic bass octaves; several accidentals were omitted (in this volume, missing accidentals are indicated by a comma: e.g. *G′,A′* signifies the absence of *G♯′*; see Chapter One, p.15; *see also* SHORT OCTAVE).

Clavecin (Fr.). Harpsichord.

Clavecin brisé (Fr.). A travelling harpsichord, made in three hinged sections, that could be folded up into a box. It was invented by Jean Marius in 1700.

Clavicembalo (It.). Harpsichord.

Clavichord [Klavichord] (Ger.; Fr. *clavicorde*, It. *clavicordo*, Lat. *clavicordium*, Port. *clavicórdio*). Clavichord.

Clavicordio (Sp.). (1) A generic term for a string keyboard instrument.

(2) Clavichord.

(3)The common Spanish term for a harpsichord.

Clavicymbalum [clavisimbalum] (Lat.). Harpsichord.

Clavier (1) (Fr.). KEYBOARD.

(2) (Ger.). A generic term for a keyboard instrument.

Claviorganum. Quasi-Latin term denoting a keyboard instrument with strings and pipes. The English equivalent is 'claviorgan'; see Chapter Four, §6.

Close plucking. *See* PLUCKING POINT.

Compass. The range of notes of the keyboard, from the lowest to the highest. It is common practice (following the system based on that of Helmholtz) to designate middle C as *c′*; the octaves above are *c″*, *c‴* and *c⁗* etc, and the octaves below *c*, *C* and *C′*. Another system uses numbers, i.e. *c‴* is written c^3, and may repeat the capital letters in the lower octaves, e.g. *C′* = *CC*. A system used in some countries (e.g. Italy) is transposed by two octaves so that *c‴* = c^5.

Contrasting double [expressive double]. A term that denotes a two-manual harpsichord in which a contrast in volume of sound or of timbre is possible by playing on either the lower or the upper manual. Such instruments have a separate set of strings for each keyboard. The term is used to differentiate such instruments from some Ruckers two-manual harpsichords, in which there is only one set of strings for the two keyboards; these Ruckers instruments are sometimes referred to as non-aligned doubles or 'transposing' harpsichords (see Chapter One, §3(i)).

Coupler. A mechanism in a two-manual harpsichord that allows the strings normally played from the upper manual to be played also from the lower manual. The term is normally applied to the 'push' or 'shove' coupler, but a DOGLEG JACK is also a form of coupler. To engage the shove coupler the upper manual is slid backwards a few millimetres so that the wooden pegs (coupler dogs) in the lower-manual keys engage the upper-manual key ends and raise them when the lower manual is played (see fig.14). Some harpsichords (mainly German ones) had dogleg jacks and a shove coupler.

Coupler dog. Part of the coupler mechanism; *see* COUPLER.

Cravo (Port.). Harpsichord. The term may also denote the clavichord.

Cross bar. A bar or rib which is glued on the underside of the soundboard for structural or acoustical reasons; it crosses the bridge (i.e. it is not glued parallel to it). In many harpsichord-making traditions, cross bars were used across the full width of the soundboard, from the spine to the bentside.

Curved tail. In a harpsichord, a tail that is combined with the bentside in a single S-shaped piece (see figs.16 and 20), forming a DOUBLE BENTSIDE. (*See also* TAIL.)

Cut-off bar. A bar which limits the area of vibration of the soundboard. It is glued to the underside of the soundboard and is usually a straight rib placed roughly parallel to the bridge, masking off (or 'cutting off') a triangular area of soundboard between the spine and the belly rail. In harpsichords without a 4′ stop there is often a cut-off bar following roughly the line of a 4′ hitch-pin rail which serves to limit the vibrating area of soundboard to a strip roughly parallel with the bentside. The 4′ HITCH-PIN RAIL is in effect also a cut-off bar.

Cut-through lute stop. *See* LUTE STOP.

Damper. The part of the mechanism that stops the vibrations of the strings when the jack returns to its rest position. A piece of woven cloth or felt is fixed on to the jack so that it either hangs on the string or wedges itself beside the string (see fig.1).

Diapason. *See* RACK.

Disposition. The registers of a harpsichord and their arrangement. For example, a single unison register is usually designated as 1 x 8′; an octave higher (the four foot) as 1 x 4′. A three-register harpsichord with two unisons and an octave would be designated 2 x 8′, 1 x 4′.

Dogleg jack. A type of jack that can be engaged by either manual of a two-manual harpsichord, thus allowing a rank to be used by both manuals. Half of the width of the jack is cut away so that its lower end reaches the lower-manual key-lever; the upper-manual key-lever engages the corner of the cut-out portion (see fig.21). It was often used in Kirckman and Shudi harpsichords.

Double bentside. (1) A term most commonly used to denote a bentside with two curves, of which one end is connected to the spine and the other to the cheekpiece, thus forming a combined bentside and tail (*see also* TAIL).

(2) A term for harpsichords made by Johan Daniel Dulcken that had a double bentside consisting of two bentsides parallel to each other, a short distance apart. The strings were attached to the outer bentside and the soundboard was glued to the inner one. In this way Dulcken presumably intended to prevent case distortion caused by string tension from having an adverse effect on the soundboard.

Double harpsichord [double; double-manual harpsichord]. A harpsichord with two manuals.

Double pinning. A system of bridge (and sometimes nut) pinning in which a second pin on the side of the bridge is so placed that in order to reach it the string has to make a small deviation after passing the bridge pin. This ensures adequate contact between the string and the bridge, enabling the string to be led towards a HITCH-PIN RAIL at the same or similar height as the bridge while avoiding excessive downward pressure on the soundboard. Usually only the bottom octave is double pinned.

Dowel. A wooden peg used for fixing in position one piece of timber with respect to another. In Ruckers instruments a square peg dipped in glue was hammered into a round hole to hold frame parts together.

Drawstop. A hand stop which is operated by being pulled out.

Eight foot [8′]. A term used in reference to organ stops, and by extension to other instruments, to indicate that they are pitched at unison or 'normal' pitch. This is not a specification of a pitch level in absolute terms (i.e. in Hz, or complete vibrations per second). For string keyboards, 8′ pitch can be assumed to be roughly a tone below modern pitch, within a margin of about a major 3rd above or below. If the pitch were to fall far outside this range then it would be designated with another pitch term: for example, Four foot (4′) pitch (an octave higher).

Enharmonic keyboard. A keyboard with more than 12 keys and sounding more than 12 different pitches in the octave (see fig.5). The extra keys make available pure intervals instead of the impure ones necessary in any tempered tunings (*see* TEMPERAMENTS). Enharmonic keyboards were made in most countries, but were more commonly used in Italy than elsewhere.

Epinette (Fr.). (1) Spinet.

(2) Virginal.

(3) A generic term for a string keyboard instrument in France until well into the 17th century.

Escapement. The part of a piano action which permits the hammer to become disengaged from the mechanism it carries towards the strings, just before the moment of impact of hammerhead and string.

Expressive double. *See* CONTRASTING DOUBLE.

False inner-outer. The term for a style of Italian harpsichord construction in which an instrument with a thick, painted cypress case appears to be in a separate outer case. It is a *trompe l'oeil* effect, in which cypress mouldings and veneer glued to the inside of the case, above the soundboard, create the appearance of a separate, removable instrument. (*See also* INNER-OUTER.)

Foreshortening. The reduction in length of the tenor and bass strings to less than the doubling theoretically required when the pitch is lowered an octave. (*See* PYTHAGOREAN SCALING and SCALE.)

Fork. The slot in a jack in which the tongue is held.

Four foot [4′]. The pitch an octave above 8′ pitch. (*See also* EIGHT FOOT.)

Four-foot hitch-pin rail. The wooden batten (liner) into which the 4′ hitch-pins are fixed. It is glued to the underside of the soundboard and against the case.

Framing. A general term for the structural members in the case of a string keyboard instrument.

Fretted. The term used for a clavichord in which more than one tangent strikes a given string (usually a pair of strings), producing two, three, or even four different notes according to the distance from the bridge, but only one at a time.

Fundamental. The lowest natural frequency of vibration of a resonator, e.g. a string (*see* HARMONICS).

Gap. The space between the belly rail and the wrest plank in which the jackslides are located.

Gauge number. The number which indicates the thickness of wire used for stringing keyboard instruments. Wire for harpsichords and other string keyboard instruments was drawn to particular sizes and each size (diameter in millimetres) was given a particular number. Sometimes the thinnest wire was numbered 11 or 12, ranging down to 1 or 0 for the thickest, but in at least one system the thickest wire had the highest number. These gauge numbers can be found in written instructions and sometimes marked on the wrest plank (or on the keys) of an instrument. They indicate the size of wire that should be used in restringing the instrument (assuming, of course, that the numbers came from the maker and not some later restorer).

Genouillère (Fr.). KNEE-LEVER.

Gravicembalo (It.). A term for the harpsichord used in Italy, principally in the 16th century. It is uncertain whether 'gravi-' is a corruption of 'clavi-', or whether it implied that the instrument was at a low pitch.

Guide. A term for the JACKSLIDE.

Half-hitch. A register giving a softer sound than normal. The jackslides are advanced only partially, so that the plectra pluck the strings less strongly.

Hand stop. A register stop operated by hand, as opposed to one operated by means of a pedal or knee lever. (*See also* DRAWSTOP.)

Harmonics. The individual pure sounds which are normally present and which together constitute what is heard as an ordinary musical note, such as from a vibrating string. The vibrations of a string are not restricted to the fundamental frequency, but are also (in theory) comprised of whole number multiples of it. Many of the harmonics theoretically possible are actually present in a musical note; they blend with the fundamental to give the impression of a single note. The fundamental (i.e. the note heard) is designated the first harmonic; the second harmonic is at a frequency twice that of the fundamental, the third at a frequency three times that of the fundamental and so on. The strength and thereby the audibility of the harmonics depend on the method of exciting string vibration (e.g. plucked or struck) and on the place along the string length where it is set in motion. In theory, if the string is plucked at a third of its length the third harmonic will be absent. (*See also* INHARMONICITY and OVERTONE.)

Harp stop. *See* BUFF STOP.

Header. *See* BELLY RAIL.

Hitch-pin. A pin of iron (or brass) to which one end of a string is attached. (*See also* TUNING PIN.)

Hitch-pin rail. A batten of wood into which the hitch-pins are driven. The 4′ rail is glued to the underside of the soundboard. Some Italian harpsichords had no 4′ rail: the pins were driven into the cypress soundboard and secured on the underside with a drop of glue. The 8′ hitch-pin rails also serve as the bentside soundboard liner and the liner glued against the tailpiece, but there is often a strip of wood glued on top of the soundboard (and against the case side) into which the pins are set; this raises the string ends and thereby decreases the downward pressure on the bridge and soundboard. In the last octave or so of many instruments it was usual to have a hitch-pin rail as high as the bridge.

Inharmonicity. A quality of timbre caused by a string's vibration not being harmonic, i.e. the overtones are not in a whole number series with the fundamental (*see* HARMONICS). Any string used in a musical instrument has some degree of inharmonicity; this is caused by the stiffness of the string, which tends to raise the frequency of the harmonics slightly.

Inner-outer. The term for the style of Italian thin-cased instrument with decorative mouldings that is kept in a separate, decorated outer case. (*See also* FALSE INNER-OUTER.)

Instrument (Ger.). A term for the virginal.

Jack. The part of the mechanism of a plucked string keyboard instrument that sets a string in motion (see fig. 1*b*).

Jack guide. *See* JACKSLIDE.

Jackrail. The padded rail above the jacks that prevents the jacks from being catapulted out of the instrument. Sometimes the jackrail also serves to limit the depth of touch of the keys (see fig. 14).

Jackslide [guide, slide]. The wooden, slotted strip which guides the jacks and holds them in position. It is either of one-piece construction (a BOXSLIDE) or two-piece. In the two-piece design the top slide (upper guide) is usually movable and the lower guide fixed (see fig. 1). In 18th-century French instruments the guides were often covered with leather, with punched slots as the bearing surface for the jacks, in order to reduce action noise. The term 'register' is sometimes used to denote a jackslide.

Japan [japanning, japanned]. A term for CHINOISERIE style decoration.

Jeu de luth (Fr.). BUFF STOP.

Just scaling. *See* PYTHAGOREAN SCALING.

Key [key-lever]. In such instruments as the harpsichord, clavichord, virginal etc, a balanced lever which when depressed by the finger operates the action, causing the string to vibrate. Key-levers can be made of one of several woods: lime, poplar, pine, beech and chestnut were all used in historical instruments. A vertical hole is drilled near the centre of the key and opened out to a slot at the top (see fig. 1*a*). A metal pin (balance pin) serves to locate the key on the balance rail and to prevent it from tipping to one side or the other while allowing the key ends to move freely up and down. Guiding of the key end is usually by a metal pin or wooden slip which engages a slot in the rack (see fig. 14). Historical string keyboard instruments (other than pianos) usually have the guiding arrangement at the end furthest from the player. The place where the jack sits on the key is padded with a piece of cloth.

Keyboard [manual] (Fr. *clavier*; Ger. *Klaviatur, Tastatur*; It. *tastiera, tastatura*). A set of levers (keys) actuating the mechanism of a musical instrument such as the harpsichord, clavichord, piano etc. The keyboard probably originated in the Greek hydraulis (the ancient pneumatic organ), but its role in antiquity and in non-European civilizations appears to have remained so limited that it may be considered as characteristic of Western music. Its influence on the development of the musical system can scarcely be overrated. The primacy of the C major scale in tonal music, for instance, is partly due to its being played on the white keys, and the 12-semitone chromatic scale which is fundamental to Western music, even in some of the more recent developments (e.g. Schoenberg's 12-note system), could be derived from the design of the keyboard. The arrangement of the keys in two rows, the sharps and flats being grouped in twos and threes in the upper row, already existed in the early 15th century.

The earliest European keyboards were simple contrivances, played with the hands rather than the fingers. Praetorius (2/1619) and some paintings of portative organs show small tabs or levers projecting from the case. Up to the 13th century the keyboards were usually diatonic except for the inclusion of B♭. By the middle of the 14th century keyboards were almost entirely chromatic. The Robertsbridge Codex (*c*1320), the earliest surviving keyboard music, requires a range of two octaves and a third, from *c* to *e''* (fully chromatic above *f*).

Before the second half of the 15th century the lowest part of keyboard compositions was often based on plainsong, or written in plainsong style. Owing to the limited number of transpositions then performed, there was no need for chromatic degrees other than the B♭ in the bass of the keyboard. This explains why the bass octave (or pedal keyboards) remained diatonic up to a late date.

The most common keyboard compass in the late 15th century and into the 16th was from *F* to *a''*, usually without *F*♯ or *G*♯. In Italy, upper limits of *c'''* or *f'''* were common in the 16th century. The lower limit was extended to *C*, usually with short octave, by the second half of the 16th century. Harpsichords with the range *F* to *f'''* were made from about 1700.

Wide keys, as in the early keyboards, suit simple and slow melodies, but make the playing of more than one part in each hand difficult. In describing key widths it is usual to measure the octave span (seven naturals) or a three-octave span (21 naturals) (*see* STICHMASS). In the 15th century octave spans were about 18 cm and in the 16th and 17th centuries were between 16.2 and 16.7 cm. Narrower spans are found (often in French instruments) of about 15.5 to 16 cm.

Keyboards may be either 'black' or 'white' according to the material covering the naturals, usually ebony, bone, or ivory. In Italy the yellowish-coloured boxwood was widely used. Where the naturals are dark the accidentals are usually light: ebony naturals with bone-topped accidentals was a common arrangement. Accidentals were sometimes made of solid ebony, but because of the high cost of this exotic wood were usually made of black-stained pear wood with a thin ebony veneer on top. Precious stones and tortoise-shell or mother-of-pearl were also used in some elaborate instruments.

Key-dip. The amount that the key can be pressed down before it comes up against its end stop or keyrail, or before a jack is arrested by the jackrail. In some Italian harpsichords, and in most virginals or spinets, the key-dip is relatively shallow – about 5 to 6 mm. Where it is necessary for three registers to pluck successively, more space is required and about 9 mm is usually provided.

Key-frame. The frame underneath the keyboard which supports the balance rail and the rack, or any other means of guiding the keys.

Keyhead. The part of the key-lever nearest to the player.

Key-lever. *See* KEY.

Key-well. The area in which the keyboard is located; it is formed by the case sides and the belly rail.

Kielflügel (Ger.). Harpsichord.

Klavecimbel (Dutch). Harpsichord.

Klaviatur (Ger.). KEYBOARD.

Knee. A triangular block glued between the baseboards and the case-side. Knees are usually found only in Italian instruments.

Knee-lever (Fr. *genouillère*). A device operated by the knee to bring into action a register or special effect stop.

Lautenzug (Ger.). BUFF STOP.

Lid swell. *See* SWELL.

Liner. The batten of wood glued to the inside of the case, to which the soundboard is attached.

Listing. Strips of cloth woven between the strings of a clavichord. When the key is released, causing the tangent to drop from the string, the cloth damps out the vibrations of the string, silencing the tone (see fig.38).

Liuto (It.). BUFF STOP.

Lower guide. A type of JACKSLIDE. It is usually fixed to the case. (See fig.1.)

Lute stop [cut-through lute stop, nasard]. A row of jacks that plucks one of the unison registers of a harpsichord very close to the nut, producing a characteristically penetrating nasal sound. The alternative 'cut-through lute stop' is derived from the gap that is cut through the wrest plank to accommodate the jacks. The German term *Nasalzug* and the French *nasale* describe more clearly than the English the character of the sound. There is some confusion as to the proper use of the term, since apparently equivalent names for the stop, such as the French *jeu de luth* and the German *Lautenzug*, actually refer to the BUFF STOP.

Machine stop. A device operated by a single pedal which controls two or more separate registers, overriding their individual handstops. It was applied to English harpsichords in the second half of the 18th century.

Manicorde [manicordion] (Fr.). (1) A term for the clavichord.

(2) A generic French term for a string keyboard instrument.

Manicordio [monacordio] (Sp.). Clavichord.

Manicordo (It.). The name used for the clavichord in Italy in the 15th and 16th centuries.

Manual. A KEYBOARD that is played by the hands (in contrast to one played by the feet, i.e. a pedal, or pedal-board).

Marquetry. A pattern or design in wood veneers which is glued to a surface as decoration.

Mensur (Ger.). *See* SCALE (1).

Mensurverlauf (Ger.). *See* SCALE (2).

Moderator. A stop that produces a softer, veiled sound in a fortepiano or tangent piano. A cloth-carrying rail is moved by a knee-lever or pedal so that the cloth is introduced between the hammers and the strings. The moderator is not identical with the 'soft' pedal on a modern pianoforte since the effect is achieved by different means.

Monochord [monochordium; canon harmonicus]. A musical device, with a single string, used mainly for teaching, tuning and experimentation. It was first mentioned in Greece in the 5th century BC and used until the 19th century. Such instruments were represented in many theoretical works of the 15th century and earlier, to demonstrate the relationship between musical intervals and corresponding string lengths. The term was used by many 15th- and 16th-century writers to denote the clavichord (see Chapter Three, §2).

Muselar [muselaar]. A term used by Claas Douwes (1699) and revived by modern writers to designate Flemish virginals which have their keyboards placed off-centre to the right and consequently have strings that are centrally plucked for most of the instrument's range. This gives the muselar a distinctive flute-like tone, unlike that produced by a virginal of any other design (see Chapter Two, §1 and fig.30).

Nag's head swell. *See* SWELL.

Nameboard. A piece of wood placed transversely above the keyboard on which the maker's name usually appears (a signature is sometimes found on the jackrail or on some part of the instrument not open to view, e.g. on the keyboard). Depending on the style of construction, the nameboard may be removable or fixed to the instrument. Sometimes it consists of two parts, one fixed and a removable, smaller, batten (on which the name may be written, as for example, with Ruckers instruments).

Nasale (Fr.; Ger. *Nasalzug*). *See* LUTE STOP.

Nasard [nasat]. *See* LUTE STOP.

Natural. On a (piano) keyboard, one of the 'white' notes. On harpsichords and other early instruments the key covers for the naturals are often of ebony, and therefore black.

Non-inner-outer. The term for a style of Italian string keyboard construction in which the case and the instrument are one and the same, without any decoration to make it appear to have a separate outer case. (*See also* FALSE INNER-OUTER and INNER-OUTER.)

Nut. In harpsichords the term that usually denotes the non-sounding bridge located on the wrest plank near the tuning pins. Some harpsichords have free soundboard wood under the nut (e.g. the 1579 instrument by Lodewijk Theewes; see Chapter One, p.15). In a virginal, if both bridges are placed on free soundboard (which is usual) the left-hand bridge (assuming the tuning pins are on the right as seen from the keyboard) is usually designated the nut. Usage on this point is not consistent.

Ottavino (It.). An octave spinet, i.e. one that plays at 4′ pitch.

Octave span. The space occupied by seven natural keys, as measured at the player's end of the key-levers. (*See also* STICHMASS.)

Overrail. A batten of wood above a specified part of the instrument. The term often denotes the rail above the key-levers, at the ends beyond the jacks, which limits the key-dip; it is found in many Flemish instruments.

Overspun string. A string in which a core is covered by a layer of wire wound uniformly around it. The earliest overspun strings, used in some large clavichords and in square pianos, did not have a close-spaced winding (i.e. one which covered the core entirely). The core is usually of iron or (later) steel wire and the over-winding of brass or copper. Overspun strings were rarely used in traditional harpsichord making but they have been used in some modern short-cased designs.

Overtone [partial]. A harmonic. The vibrations at twice the frequency of the fundamental note are known as the first overtone, or as the second harmonic (*see* HARMONICS).

Pantalon stop (Ger. *Pantalonzug*, *Pantaleonzug*). A device found in a few unfretted clavichords of the second half of the 18th century. It was named after the Pantaleon (or Pantalon), a large dulcimer invented by Pantaleon Hebenstreit (1667–1750), the tone of which was enhanced by the resonance of undamped strings. These strings had their own, separate dampers. (See Chapter Three, p.166.)

Partial. *See* OVERTONE.

Peau de buffle (Fr.: 'buffalo hide'). The term for a register found in late 18th-century French harpsichords in which the jacks had buffalo-hide plectra instead of the usual quill. It should not be confused with the hard sole leather used for plectra in some modern harpsichords.

Pedal. (1) A lever, operated by the foot, which controls a register or effect on a harpsichord.

(2) The term (also 'pedal-board', both derived from the organ) for a series of pedals, arranged somewhat like the keys of a piano or harpsichord (i.e. with different levels for naturals and sharps), to form a

keyboard played by the feet. Pedal keyboards were sometimes provided for the harpsichord and clavichord and operated an independent instrument underneath the main one. An arrangement quite often used in Italian harpsichords was for a pedal-board to cover the range *C/E* to *c* (eight notes). This arrangement, known as a pull-down pedal, was without its own sounding instrument: the pedals were attached by cords to the key-levers of the harpsichord so that notes could be held or played from the pedal-board as on an Italian organ.

Pin-barrel mechanism. A mechanism consisting of a barrel into which pins are fixed. The barrel rotates slowly and the pins produce a note or control a function. It is found in many automatic instruments; the commonest is the musical box, which has pins that pluck small strips of metal to produce the sounds. Some late 16th-century German spinets were equipped with a pin-barrel mechanism.

Pin block. The American term for WREST PLANK.

Pivot. *See* AXLE.

Plectrum. The piece of quill or plastic projecting from the tongue of a jack which plucks the string. In 18th-century French harpsichords buffalo leather was sometimes used instead of quill, and brass plectra were used in some Renaissance harpsichords. There is no historical justification for the modern use of hard sole leather in old instruments.

Plein jeu (Fr.: 'full registration'). The fullest registration of which an instrument is capable.

Plucking point. The place on a string, relative to the nut, where it is plucked. The closer the plucking point is to the nut ('close plucking'), the more nasal the tone, since the higher harmonics are highly excited; the closer it is to the centre of the string ('centre plucking'), the rounder the tone sounds. Thus the plucking point is one of the most important determinants of the sound of a plucked string keyboard instrument. For practical reasons the plucking point remains fixed, so that it is not possible to vary the tone by varying this parameter. (See also Chapter One, p.13.)

Polygonal spinet. A term that usually denotes a five- or six-sided plucked string keyboard instrument. The name 'polygonal virginal' is usually preferred for this type of instrument, but since German uses 'Spinett' and Italian 'spinetta', no consistent usage has developed.

Polygonal virginal. *See* POLYGONAL SPINET.

Pull-down pedal. *See* PEDAL (2).

Push coupler. *See* COUPLER.

Pythagorean scaling [just scaling]. Scaling design which follows the principle of halving or doubling the length of a string at each octave. Theory requires, other factors being unchanged, that when the pitch is halved, the length of a string must double. Pythagorean scaling is not practical throughout the range of a normal-sized harpsichord since the bass strings would be several metres long; the scaling is therefore reduced, or foreshortened, by increasing the tension (*see* SCALE).

Querspinett [Querflügel] (Ger.). *See* BENTSIDE SPINET.

Quilling. The practice of inserting and trimming plectra; these were usually of bird quill (e.g. raven). In order to achieve a light and reliable touch the plectrum must be of such a length that it does not project past the string too much. The underside of the plectrum is scraped or pared with a sharp knife until the required strength of pluck has been achieved. In modern practice plastic is widely used. (*See also* PLECTRUM.)

Quint ('fifth'). A term that usually denotes a pitch a 5th above 8′ pitch in string keyboard instruments.

Rack [diapason]. The batten of wood that guides the jack ends of the key-levers. It is fixed to the keyboard frame and has vertical slots in it. A metal pin or thin wooden tongue in the jack end of the key-lever engages a slot in the rack (see fig.14).

Rank. (1) A row or line of jacks.
(2) A term for REGISTER.

Ravalement (Fr.). A term for the alteration and extension of the range of keyboard instruments. It is most often applied to the rebuilding of Ruckers instruments that was widely practised in Paris in the 18th century; rebuilt (*ravalé*) Ruckers harpsichords fetched high prices. A harpsichord with a range of *C*/*E* to *c‴* might have been modified to make the compass chromatic from *C* to *d‴*; this could have been carried out within the existing case width by making a new keyboard and jackslides and repinning the bridges. This would be termed a small (*petit*) *ravalement*; a large (*grand*) *ravalement* involved widening the case. The type of modification depended on the prevailing musical requirements in the country in which it was carried out. Thus an English modification would not necessarily have been identical with a French *ravalement*.

Register. A term synonymous with STOP: for example, an 8′ register means 8′ stop. It is also sometimes used to refer to a JACKSLIDE.

Ribbing. The system or arrangement of the small strips (ribs) of wood that are glued to the underside of a soundboard for acoustical and structural reasons. Ribs are usually deeper than they are wide. (*See also* BARRING.)

Rollerboard. A system of intermediate levers which enables a key to move a part of the action that is not immediately above or beside the key-lever. The rollerboard is part of the tracker action of an organ, but it is also used in a type of clavicytherium that has a symmetrical ('pyramid') case, which requires that the bass strings are in the centre of the instrument and the treble strings on the outside. The rollerboard is employed to connect each key with its appropriate jack and string. The use of a rollerboard can be avoided, however, if diagonal stringing is used.

Rose. The decorative device in the soundboard of a keyboard instrument. It is usually circular and either carved from the wood of the soundboard itself, as in some early Italian harpsichords and virginals, or, most often, set into the soundboard. In the Italian tradition layers of wood veneer or parchment were used to build up a design. In Flemish, and other, traditions a rose was cast from lead (or lead alloy). Ruckers instruments display an angel playing a harp and the maker's initials (see fig.11; see also figs.6, 22, 25, 30, 32, 36, 47).

Scale [scaling]. (1) (Ger. *Mensur*) The string length at a given note. It is common to give the string length of the note *c''* as an indication of the pitch of a string keyboard instrument.

(2) (Ger. *Mensurverlauf*) The sounding length of the string in relation to the intended pitch. Given constant tension and diameter, a string sounding an octave below another must be twice as long. Since this doubling would result in an impracticably long instrument, the strings are made shorter than the theoretical length and string tension is increased to compensate. Thus the scale is said to have been shortened. (*See also* FORESHORTENING and PYTHAGOREAN SCALING.)

Scroll sawn. A term for the decoration involving flowing curves at the edge of the cheekpiece in Italian harpsichords and virginals.

Short octave (Fr. *octave courte*; Ger. *kurze Oktave*). A term to denote the tuning of some of the lowest notes of keyboard instruments to pitches below their apparent ones, a practice employed from the 16th century to the early 19th to extend the keyboard compass downwards without increasing the overall dimensions of the instrument. The short octave was basically a variable tuning adapted to the requirements of individual pieces. It was first applied to keyboards showing *F* as the lowest key; the *F*♯ and *G*♯ keys, if present, were tuned to sound lower notes, usually *C*, *D* or *E*. By the middle of the 16th century, an apparent *E* was added as the lowest key, but it was often tuned to a lower pitch. This soon resulted in the standard tuning known today as the '*C*/*E* short octave'. From the 17th century onwards composers often demanded a chromatic compass and so keyboards were enlarged (*see* RAVALEMENT); or else the two lowest upper keys were split into two parts, the front tuned to the short octave note, and the back to its proper note (*see* BROKEN OCTAVE).

Short scale. A short-scaled instrument is one which has a *c''* string of about 25 to 28 cm; by contrast, a long-scaled instrument would be about 32 to 36 cm at *c''*. These scales are intended for normal 8′ pitch; they do not represent different pitches, nor widely differing string tension: the difference is due to the use of different string materials. Thus the short scales are intended for brass wire strings, the long scales for iron wire. (A short scale of 28 cm strung in brass would come to about a semitone below modern pitch, in which *a′* = 440 Hz.)

Shove coupler. *See* COUPLER.

Slide. *See* JACKSLIDE.

Sordino (It.). A term for the clavichord; it is no longer used.

Soundboard. The thin sheet of wood that serves to make the vibrations of the strings more audible and helps to form the characteristic tone quality of a string keyboard instrument. A string presents so small a surface to the surrounding air that its vibrations cannot set the air in motion with any great efficiency. However, the soundboard does not act as an amplifier in the same sense as an electronic circuit since it contributes no energy to the vibrations. Rather, it enables the energy already imparted to the string by a hammer, plectrum or tangent to be converted more efficiently into vibrations of the air. The particular resonance and vibrational characteristics of the soundboard determine which components of the complex vibrations of the string will be given particular prominence. Consequently, the shape, thickness and ribbing of the soundboard are of primary importance in determining the quality of the instrument. Normally, quarter-sawn spruce (*picea excelsa* or *abies*) is used since its ratio of stiffness to weight is favourable, but some instruments have soundboards of fir (*abies alba*). (Spruce and fir can be reliably distinguished only by microscopic examination.) Many Italian instruments had cypress (*cupressus sempervirens*) soundboards; maple was also occasionally used.

Soundhole. An opening made into the volume underneath the soundboard. A soundhole in the soundboard is usually circular, with a decorative ROSE. In a clavichord, a soundhole is often made in the cross-brace under the left-hand edge of the soundboard; it may not be visible, since it is below the level of the keys, and has no rose.

Spine. The long, straight side of a harpsichord, next to the bass strings, opposite the bentside.

Spinett (Ger.). German term for the spinet. It was (and sometimes still is) also used for instruments which are given other names in current organological terminology: e.g. 'polygonal virginal' is 'polygonales Spinett'. The large Ruckers virginals with the keyboard placed towards the left-hand side of the instrument were called 'Spinetten'.

Spinetta [spinetto] (It.). A name used occasionally in the 16th century in Italy, and widely thereafter, for the polygonal virginal. The diminutive 'spinettina' ('spinettino', also 'spinetta ottavina') denotes an instrument at 4′ pitch.

Spinettone (It.). A large SPINETTA. Cristofori made an instrument described as 'spinettone da orchestra'; this instrument has been described elsewhere as a CEMBALO TRAVERSO.

Split sharp. A sharp key divided or 'split' into two parts: the front part is about one third of the length of the whole. Usually the back part is set slightly higher to facilitate playing. Each part has its own key-lever, jacks and strings so that two notes are available. In Italian instruments it was common – and to a lesser extent in Flemish and German instruments – to provide split sharps for *e♭/d♯* and *g♯/a♭*. The usual practice was to put on the front part the note that would normally be found there, e.g. *e♭* and *g♯*.

Spring. The part of the jack mechanism that biases the tongue to its forward position, and allows it to swing back as the plectrum passes the string on its downward path after having plucked the string (see fig. 1*b* and *c*). It was usually made of pigs' bristle, but Italian makers used a thin, flat, brass-leaf spring and modern springs may be of plastic. Its adjustment is critical for the correct functioning of the action: if it is too strong the plectrum may hang on the string and will therefore not be ready to pluck a second time. If too weak, the repetition will be poor, or the plectrum may not touch the string at all since the tongue will not be biased properly to its forward position.

Standard measurement. *See* STICHMASS.

Stepped nut. A one-piece nut which has two different levels for strings. The strings on the lower level pass through holes in the nut to reach the tuning pins.

Stichmass (Ger.). A term for the measurement of the size of a keyboard. Instead of an octave (seven natural keys), the standard measurement is of three octaves (21 keys). This larger measurement has the slight advantage of evening out any inconsistencies in the keyboard due to manufacture or distortion with time.

Stop [register]. A rank of strings. 'Stop' is commonly used for any register that provides a special effect (e.g. BUFF STOP) and the device by which it is operated.

Stop knob. A knob which operates a register. It may be pushed and pulled or slid from side to side for engagement and disengagement of the register.

Stretcher. A frame member or part of a stand which connects two parts. The term is frequently used for the pieces of a stand which join the legs together.

String. In a musical instrument, a length of any material that can produce a musical sound when held under tension and plucked, bowed, struck or otherwise excited. Wire, gut and silk are the most commonly used materials; metal wire is used almost exclusively for string keyboard instruments. In theory, the frequency at which a string will vibrate is inversely proportional to its length, proportional to the square root of its tension and inversely proportional to the square root of its mass. In theoretically free vibration (i.e. vibration not hindered by damping or stiffness in the string, not damped by the air or by attached bodies such as bridges and soundboards) a string will produce harmonic overtones, i.e. overtones which are whole-number multiples of the basic frequency (the fundamental). In practice there are several factors which cause the overtones to depart from the harmonic series and cause inharmonicity. The inharmonicity caused by string stiffness is relatively unimportant as a factor determining timbre in plucked string keyboard instruments (it is more important in instruments such as the piano). Although the string is the immediate source of energy (in practical terms, the player's finger is the ultimate source which sets the string and thereby the soundboard in motion), it must not be imputed to the string that which is really the contribution of the soundboard, or the result of the interaction of the string and soundboard. In other words, strings alone do not determine timbre.

Iron, yellow brass (75–70% zinc, 25–30% copper), red brass (90–85% zinc, 10–15% copper) and occasionally pure copper strings were all traditionally used for string keyboard instruments. Iron wire was a hard-drawn, virtually pure iron and only in the 19th century were steels, which have a higher, strengthening, carbon content and consequently higher tensile strengths, widely available. The scalings of tenor and bass strings were designed to match the tensile strengths of the available wires. Thus harpsichords with a scale of about 35 cm at *c''* designed for iron stringing in the treble, would change to yellow brass wire at about *c*, and to red brass for the last few bass notes. Short-scaled instruments (*c''* of about 25–28 cm) were intended to be strung with yellow brass wire, possibly changing in the bass to red brass.

Stringband. The strings and the space occupied by them, measured across the instrument.

Strut. A part of the framing that supports the case. Struts are most important in maintaining the shape of the bentside, to prevent collapse or distortion.

Swell. A device that enables the volume of sound to be varied. Such devices were applied to English harpsichords (and some square pianos) in the last quarter of the 18th century. In the earlier version, a 'lid swell' or 'nag's head swell', a pedal was depressed to raise gradually a hinged section at the right-hand side of the harpsichord's lid. In the second type, the 'Venetian swell', the soundboard was covered by an inner lid fitted with pivoted louvres, like those of a window blind, which could be operated by a pedal. The lid swell is first mentioned in the patent specification of Roger Plenius's 'lyrichord' (1741) and seems to have been applied to harpsichords in the early 1760s. The Venetian swell was patented by Burkat Shudi in 1769 and seems to have been an improvement only to the extent that the operation of its louvres is visually less obtrusive than the flapping of a large section of the instrument's lid.

Tactus (Lat.). The term used by the Italian scholar Giorgio Anselmi (*De musica*, 1434) for a key on a clavichord or organ.

Tail. The short side of a harpsichord or bentside spinet, furthest from the player, between the straight long side (spine) and the bentside. In some instruments (especially German) the tail is formed by an additional, outward, curve in the bentside, in place of the normal straight piece with a joint (see fig. 20); there is no special advantage in this arrangement. (*See also* DOUBLE BENTSIDE.)

Tangent. The part of the action of a clavichord which strikes the string. It is usually made of flat brass sheet (about 1 mm thick) and set into the key-lever.

Tangentenflügel (Ger.). Tangent piano (see Chapter Four, §9).

Tastatur (Ger.; It. *tastatura*, *tastiera*). KEYBOARD.

Temperaments. Tunings of the scale in which some or all of the intervals are made slightly impure in order that few or none will be left distastefully so. Equal temperament, in which the octave is divided into 12 equal semitones, is the standard modern western temperament, except among specialists in Renaissance or Baroque music; for this music mean-tone or irregular temperaments are widely preferred.

A mean-tone temperament in its most restricted sense refers to a tuning with pure major 3rds divided into two equal whole tones (hence 'mean'-tone). A broader use of the term includes Renaissance or Baroque tunings in which a major 3rd slightly larger (or, less often, slightly smaller) than pure is divided into equal whole tones. The 12-note scale will include one sour 'wolf' 5th, considerably larger than pure because the other 11 are tempered more than enough to make a 'circle' of identical 5ths (as in equal temperament). Hence the tuner of a mean-tone temperament must choose not only a particular shade of mean-tone tempering of the 5ths (e.g. ⅓–⅕ comma), but also the position of the wolf 5th. A commonly used mean-tone tuning, usually attributed to Pietro Aaron (*Thoscanello de la musica*, Venice, 1523/*R*1969; Eng. trans., 1970), has ¼ comma tempering of the 5ths. The wolf 5th usually lies between *G*♯ and *E*♭.

Irregular temperaments, using 5ths of different sizes, were in use from the earliest times: Arnolt Schlick (*Spiegel der Orgelmacher und Organisten*, Speyer, 1511/*R*1959; Eng. trans., 1978) published a description of a tuning which was an artful variant of regular mean-tone with major 3rds slightly larger than pure. The most characteristic type of 18th-century keyboard tuning was an irregular temperament with no wolf 5th, but with the 3rds in the C major scale tempered slightly as in some forms of mean-tone temperament. Moving around the circle of 5ths, the amount of tempering of the 3rds increased, the exact pattern of the differences varying according to taste. Some German writers characterized as 'good' any temperament without a wolf 5th (thereby including equal temperament), but others regarded equal temperament as distinctly less preferable than irregular temperaments. It is a mistaken assumption that 'well-tempered' in Bach's *Das wohltemperirte Clavier* refers to a specific temperament, but it is in any event unlikely that equal temperament was intended.

Timbre. The characteristic sound quality of an instrument (*see* TRANSIENTS).

Tongue. The part of a jack that holds the plectrum (see fig. 1*b*).

Touche (Fr.). KEY.

Touch depth. *See* KEY-DIP.

Transients. The rapid changes of sound as it develops or decays in the first or last few fractions of a second. Starting transients are important for the perception of the timbre (i.e. the characteristic sound) of an instrument; if they are removed from a recording, a listener may fail to identify the instrument.

Transposing keyboard. A keyboard that enables the performer readily to play music in a different key from that in which it is written, generally for the purpose of enabling the music to sound at a different pitch or to permit the playing of music in a 'difficult' key while using the fingering of an 'easy' key. Most early transposing keyboards simply slide sideways beneath the jacks, hammers, stickers, etc, of the instrument of which they are a part. Shifting keyboards were not commonly made in historical harpsichords, although they are often included in modern reproductions. The Hans Müller harpsichord of 1537 (see Chapter One, §2 (ii)) was equipped with a transposing keyboard for a whole tone. The two-manual harpsichords made by the Ruckers family are often referred to as 'transposing' harpsichords (see Chapter One, §3(i)). In these instruments the upper keyboard was pitched a 4th higher than the lower one. It has been argued that many harpsichords and virginals in Italy were built to act as transposing instruments. Although the arguments as originally proposed are incorrect, there were nevertheless instruments made at specific pitches, which might have facilitated some transpositions (see Chapter One, §2(i)).

Transposition. The notation or performance of music at a pitch different from that in which it was originally conceived, by raising or lowering the notes in it by the same interval. Some keyboard instruments were fitted with aids to make performance at a different pitch easier. (*See* TRANSPOSING KEYBOARD.)

Trefoil. The characteristic three-lobed gothic ornament incorporated in a half circle. It is found on the fronts of the natural keys, mainly in 17th-century French harpsichords.

Tuning pin [wrest pin]. The round metal pin set in the wrest plank to which one end of a string is attached; it is rotated to tune the string. In historical instruments the tuning pin was usually about 4 mm in diameter, but 4′ registers often had slightly thinner pins in order to facilitate tuning. Modern zither or piano pins have a hole to secure the wire, but historical tuning pins had the wire simply wrapped on in a self-locking way. This is quite easy with the small diameters and relatively soft wire used for old instruments. Many old tuning pins were tapered on the last few millimetres so that they could be tapped down into the wrest plank to tighten their grip, in the same manner as violin pegs.

Una corda (It.: 'one string'). A stop used in some early pianos to produce softer than normal sound. It causes the key-frame to be moved sideways so that not all the strings available for each note are struck: one string is struck in BICHORD-STRUNG notes, two in trichord.

Unison. 8′ pitch or a stop at that pitch. (*See* EIGHT FOOT.)

Upper guide. A JACKSLIDE. Where the jackslides are of two parts, the upper guide is arranged to be movable so that the register can be turned on or off.

Venetian swell. *See* SWELL.

Virginal (Ger.; Fr. and It. *virginale*). Virginal.

Vis-à-vis (Fr.). A term applied to a type of keyboard instrument with (usually) two keyboards, one at each end (see Chapter Four, §5).

Voicing. The means by which the timbre, attack, loudness, etc, of the pipes or strings of keyboard instruments are given their desired quality and uniformity (*see* QUILLING).

Wirbel (Ger.). TUNING PIN.

Wolf. The name given to the undesirable sound effect caused by a 5th that is larger than pure (*see* TEMPERAMENTS). The term can also refer to any false-sounding notes, especially in the cello.

Wrest pin. *See* TUNING PIN.

Wrest plank [pin block]. The piece of wood into which the tuning pins are set. It is glued to the case sides at each end and to the wrest-plank blocks, which take the main pull of the strings. It is often made of oak or beech, sometimes of pear or walnut.

Wrest-plank blocks. The blocks of wood glued to the bottom boards on to which the wrest plank is glued. In some of the earliest instruments the wrest plank was simply glued to the case sides, a practice that was possible only where there was a single register and light stringing.

APPENDIX TWO

Index of Instrument Makers

Cross-references within this appendix are distinguished by the use of small capitals, with a large capital for the initial letter of the entry referred to, for example: *See* SMITH, JOHN; or, in running prose, JOHN SMITH.

Adlam, Derek (Leslie) (*b* 1938). English maker of fortepianos, clavichords, harpsichords and virginals. After studying piano, harpsichord and organ at the Guildhall School of Music he restored instruments as well as teaching music. From 1963 to 1973 he was curator of the Colt Clavier Collection, Bethersden, Kent, and in 1971 formed a partnership with Richard Burnett to produce keyboard instruments. He has worked independently since 1979. He has contributed on the subject of harpsichord restoration to *Early Music* (1976).

Antegnati. Italian family of organ makers. They were active from the late 15th century to the second half of the 17th century in Brescia. Giovanni Francesco (*fl* 1533–44) also made harpsichords and arpicordi. His 1537 polygonal virginal (in the Victoria and Albert Museum, London) is an important document for the development of instrument making in Italy.

Bach, Johann Nikolaus (1669–1753). German organist and part-time organ and harpsichord maker, related to J. S. Bach. He is known to have made gut-strung harpsichords (*Lautenklaviere*).(See Chapter Four, §7).

Backers, Americus (*fl* 1763–78). Dutch maker of harpsichords and pianos, active in London. Broadwood attributed to him the invention (by 1772) of the English grand-piano action. His only surviving harpsichord (1766) has a five-octave compass and 2 × 8′, 1 × 4′ disposition, with a pedal-operated lute stop.

Baffo, Ioannes Antonius (*fl* 1574–9). Italian harpsichord maker, active in Venice. Only two harpsichords and a virginal attributed to him are genuine; the others have faked inscriptions. The harpsichords (1574, in the Victoria and Albert Museum, London, and 1579, Paris Conservatoire) are fine examples of his workmanship.

Barnes, John Robert (*b* 1928). English instrument maker and restorer. He studied physics at the University of London and was an audio engineer before working as a professional instrument maker. He restored instruments in the Victoria and Albert Museum and the Royal College of Music, London, and published several important articles on the history of Italian keyboard instruments (1965, 1966, 1968, 1973). He was curator of the Russell Collection of Harpsichords and Clavichords, Edinburgh, from 1968 to 1983. He was associated with the development of Zuckermann instrument kits as well as building instruments. His book *Making a Spinet by Traditional Methods* (1985) deals with the constructional details of Stephen Keene's instruments. Early pianos form an important part of his private collection.

Bédard, Hubert (François) (*b* 1933). Canadian harpsichord maker and harpsichordist. He played piano and organ from an early age, before studying medicine and then music; his teachers included Kenneth Gilbert and Gustav Leonhardt. After a period in the workshop of Frank Hubbard, in 1968 he moved to Paris as chief restorer in the Conservatoire workshop. He established his own workshop and also developed kits which are marketed by Heugel in Paris.

Belt, Philip (Ralph) (*b* 1927). American fortepiano maker. He worked first as a piano technician. He made his first replica of a fortepiano (after J. L. Dulcken) while apprenticed to Frank Hubbard (1965–7). In 1966 he established his own workshop and in 1975–9 was also associated with Zuckermann Harpsichords. After restoring a Stein instrument he produced replicas and a Stein fortepiano kit (1972). His instruments have a clear sound and a good touch. He has contributed articles on early pianos and their makers to *The New Grove Dictionary of Musical Instruments* (1984).

Bertolotti. Italian harpsichord maker in Venice. A pentagonal virginal by him is dated 1585. The harpsichord apparently by him (also 1585; in the Brussels Museum of Musical Instruments) has had its nameboard altered to show his name.

Biest, Marten van der (*fl* 1557–84). Flemish harpsichord maker. He entered the Guild of St Luke as an instrument maker in 1557. Only one of his instruments has survived: a double virginal (mother and child) of 1580, now in the Germanisches Nationalmuseum, Nuremberg.

Blanchet. French family of harpsichord and piano makers, active in Paris from the end of the 17th century to the middle of the 19th. Nicolas Blanchet (*c*1660–1731) was admitted to the guild as a master in 1689. His second son, François Etienne (*c*1700–1761), became his partner in 1722. François married Elisabeth Gobin, who brought as part of her dowry a house in the rue de la Verrerie, to which the business moved. Their son François Etienne (*c*1730–1766) continued the family business, but on his death it was taken over by his chief workman, PASCAL TASKIN, as his own son, Armand François Nicolas (1763–1818), was only three years old. Armand worked with Taskin and wrote *Méthode* *abrégée pour accorder le clavecin et le piano* (Paris, 1797–1800/*R*1976). Armand's son Nicolas managed the workshop after his father's death in 1818 and later formed a partnership with Jean (Johann) Roller, a German piano maker active in Paris. When Roller retired in 1851 he was replaced by Nicolas's son, P. A. C. Blanchet, who succeeded his father in 1855. Inventories and guild records indicate that the Blanchets enjoyed financial and artistic success from the beginning. Their instruments commanded considerably higher prices than those of other makers in Paris and in the 1750s their firm was appointed 'facteur des clavessins du Roi'. They were also renowned for their reworking of 17th-century Flemish instruments. In the 19th century the firm was active in piano making.

Blyth, Samuel (baptized 1744; *d* 1795). American craftsman and organist. Apart from his activities as a painter and gilder and maker of Venetian blinds, he made musical instruments. Only one spinet is known (*c*1785; in the Essex Institute, Salem, Mass.). It is one of the few extant examples of American 18th-century string keyboard instruments. The range is G'/B' to f''' and it has a mahogany case.

Boni, Giovanni Battista (*fl* 1617–19). Italian harpsichord maker. Two of his instruments, a harpsichord of 1617 and a virginal of 1619, have split sharps.

Bridge [Bridges], **Richard** (*d* 1758). English builder of organs, harpsichords and spinets. Although his work on organs is well documented, only one spinet survives. This instrument has a five-octave compass, G' to g''', and an intricately inlaid case.

Broadwood. English firm of piano makers. John Broadwood (1732–1812) was a joiner and cabinet maker who went to London in 1761 and worked with Burkat Shudi. In 1769 he married Shudi's daughter and in 1770 became his partner. From that time the joint production of harpsichords was signed 'Shudi and Broadwood'. In the 1770s Broadwood developed the English grand piano action, but continued to make harpsichords with the 'Venetian swell' action (for which invention he paid Shudi royalties) after Shudi's retirement. By 1795, when Broadwood took his son James Shudi Broadwood into partnership, the business was wholly devoted to the pianoforte.

Bull, Joannes Petrus [Bohll, Pierre] (1723–1804). Flemish harpsichord maker of German origin. He arrived in Antwerp in 1745 and learnt harpsichord making with Dulcken. Four of his instruments survive, from 1776–89. They have a five-octave range, one or two manuals, three stops ($2 \times 8'$, $1 \times 4'$) and knee-levers operating a dogleg jack register and buff stop.

Catlin, George (1777 or 1778–1852). American instrument maker. He made a wide range of instruments, including woodwind and string instruments, as well as harpsichords, pianos and organs, besides measuring and surveying instruments. About 1810 he invented a bass clarinet, the 'clarion', which was the most successful instrument of its type until about 1830.

Celestini, Ioannes (*fl* 1587–1610). Italian harpsichord maker. He worked in Venice. He made polygonal virginals, harpsichords and an 'arcispineta' (a rectangular virginal; see p.133). Ten of his instruments have survived.

Challis, John (1907–74). American harpsichord maker. He studied organ with Frederick Alexander before going to England in 1926 to work with Arnold Dolmetsch. He was awarded the first Dolmetsch Foundation scholarship (1928). In 1930 he returned to the USA, eventually establishing his workshop in New York. His harpsichords were highly innovative and skilfully made, but their tonal conception was not in the historical tradition. Instead he sought, with the use of metals and plastics, to make instruments that would be as stable as the piano.

Colonna, Fabio [Linceo] (*c*1580–*c*1650). Inventor of an enharmonic instrument with eight keyboards. The 'Sambuca lincea' is described in Colonna's *La Sambuca lincea overo del'istromento musico perfetto* (Naples, 1618); since it has a tangent action, it is an enharmonic clavichord.

Couchet. Flemish family of virginal and harpsichord makers. Joannes Couchet (1615–55) was a grandson of Hans Ruckers (see Chapter One, §3(i)) and took over the Ruckers workshop in 1642; three of his seven children became harpsichord makers: Petrus Joannes, Joseph(us) Joannes, and Maria Abraham. Joannes was apprenticed in 1626 to his uncle Joannes Ruckers and his instruments are closely similar in construction, decoration and sound to other members of the Ruckers family. However, Joannes Couchet extended the range of instruments offered: he made a double-harpsichord with an unusual, extended upper-manual compass, and a single-manual harpsichord with 2 × 8′ (although he admitted to Constantijn Huygens in a letter that he preferred the sound of a single unison with an octave). Three double harpsichords, one single-manual harpsichord and a muselar are known to have been made by Joannes between 1645 and 1652; one of these, a double, is undated. Five harpsichords are attributed to the sons Petrus and Joseph. Of these, one is of doubtful authenticity; another has two rows of 8′ jacks which pluck one set of 8′ strings.

Cristofori, Bartolomeo (1655–1731). Italian keyboard instrument maker and inventor. He served at the Florentine court of prince Ferdinand de' Medici from 1690 and shortly after this date he must have begun work on a piano escapement, the invention for which he is chiefly remembered. An inventory of 1700 establishes that he had already completed one 'arpicembalo', an instrument with a hammer mechanism as well as a rank of jacks. Three pianos survive, made in

1720, 1722 and 1726; they are bichord-strung with a range *C* to *c'''*. Cristofori also made unusual harpsichords with the disposition 1 × 8', 1 × 4', 1 × 2' and a symmetrically-shaped 'spinetta ovale', as well as undertaking repair and rebuilding work in the prince's collection.

Delin, Albert (*fl* 1750–70). Harpsichord maker, active in Tournai (now Belgium). He is an important representative of the later Flemish style of instrument making. His output included harpsichords, spinets and clavicytheria. Some of his instruments (e.g. a harpsichord of 1750) have a relatively small range for this late date: *C* to *c'''*, with 2 × 8' registers. (See also Chapter Four, §4.)

Denis. French family of instrument makers. Its members include Robert (*d* 1588 or 1589); his sons Claude (1544–87) and Jehan (1549–after 1589); G. Denis (*fl* 1634), who was perhaps a son of Jehan; Thomas (*d* before 1620) and Pierre (*fl* 1634–42), sons of G. Denis; Jean (*d* 1672) and his brother Philippe (*fl* 1672); and Jean's son Louis (1635–*c*1711). Robert worked in Paris from 1544 as an organ and spinet builder and probably also made harpsichords. His son Claude made various string instruments, including violins and citterns, and had over 600 instruments in his workshop at the time of his death. Jean is the best-known. His spinets were praised by Mersenne. He became known through his *Traité de l'accord de l'espinette avec la comparaison de son clavier avec la musique vocale* (Paris, 1643, 2/1650/*R*1969). A spinet made by his brother Philippe is in the Paris Conservatoire. Louis Denis is the last member of the family who can be identified with certainty.

Dolmetsch. English family (of mixed French, German, Swiss and Bohemian origins) of instrument makers, scholars and performers of early music. (Eugène) Arnold Dolmetsch (1858–1940) was born in France. He learnt piano making in his father's workshop and organ building from his maternal grandfather. He studied the violin in Brussels and London and taught violin at Dulwich College (1885–9). He made his first clavichord in 1894 and his first harpsichord two years later. He worked for Chickering & Sons in the USA, producing early instruments, then with Gaveau in Paris for three years (until 1914) and in 1917 moved to Haslemere, where he established a workshop. He had a great influence on late 19th- and early 20th-century attitudes to scholarship and practice, particularly through the construction of then obsolete instruments, including viols and, notably, recorders; his keyboard instruments were not closely modelled on historical instruments. His son Carl (Frederick) (*b* 1911), best known as a virtuoso recorder player, became supervisor of the workshop. Between 1978 and 1982 the firm was split into Arnold Dolmetsch Ltd (not run by the family) and a group of other firms. In 1982 Carl and his daughters Jeanne and Marguerite (both *b* 1942) resumed control of the entire group.

Dominicus Pisaurensis [Domenico da Pesaro] (*fl* 1533–75). Italian instrument maker, apparently active in Venice. He is not to be confused with Domenicus Venetus. 'Pisaurensis' denotes that he came from Pesaro on the Adriatic coast, but this gives no indication of whether he moved to Venice, or whether his family had been there for some time. More of his instruments have survived than from any other 16th-century harpsichord maker. Although several instruments attributed to him have faked inscriptions, six harpsichords and six polygonal virginals have been authenticated and his clavichord of 1543 is the earliest dated surviving clavichord (see fig. 42 and Chapter Three, §2).

Dowd, William (Richmond) (*b* 1922). American harpsichord maker. After studying English at Harvard he became increasingly interested in early keyboard instruments and, with Frank Hubbard, constructed a clavichord. Dowd served an apprenticeship with John Challis in Detroit and in 1949 established a workshop with Hubbard in Boston, Mass. By 1955 the firm had constructed 13 harpsichords and four clavichords, besides completing several restorations. Dowd evolved a design based on the two-manual harpsichord of Pascal Taskin which soon found wide favour with performers as a general-purpose concert instrument. After the dissolution of the partnership in 1958, Dowd established his own workshop in Cambridge, Mass., attaining an annual production of 20 to 22 instruments. Between 1971 and 1985 he ran an additional workshop in Paris with Reinhard von Nagel. Dowd harpsichords are probably in wider use by leading professional performers in North America and Europe than those of any other maker.

Dufour, Claude (*c*1628–1709). French harpsichord maker. He worked in Lyon. Two of his instruments have survived: a harpsichord of 1674 and a spinet. He may be related to Nicholas Dufour, whose name appears on the label of a harpsichord dated 1683 (assuming that the label is genuine).

Dulcken. Flemish family of harpsichord makers, of German origin. Joannes Daniel (*d* 1757) made both single- and double-manual instruments, generally with a five-octave compass and 2 × 8′, 1 × 4′ disposition. Burney records that, after the Ruckers family, the 'harpsichord maker of the greatest eminence . . . was J. D. Dulcken'. Eight harpsichords by him are known to survive. His son Johan Lodewijk (Louis) (1736–after 1793) established his workshop in Amsterdam in 1755 after his apprenticeship with his father. Later he went to Paris and built pianos. Johan's brother Joannes (1742–75) is known to have made harpsichords in Brussels dated 1764 and 1769. Johan's son, also named Johan Lodewijk (baptized 1761; *d* after 1835) established himself in Munich as Mechanischer Hofklaviermacher.

Dumont, Nicolas (*fl* 1673–1710). French harpsichord maker. Active in Paris, he was admitted to the Paris guild of instrument makers in 1675. Three of his harpsichords are known (see Chapter One, §4(i)).

Elsche, Jacobus [Jakob] **Van den** (*c*1689–1772). Flemish harpsichord maker. He is the only known maker active in Antwerp between the time of the Ruckers–Couchet workshop and the arrival of J. D. Dulcken in 1738. Only one instrument survives: a two-manual harpsichord with the disposition 2 × 8′, 1 × 4′, nasard stop and two knee-levers.

Faby, P. (*fl* 1677–91). Italian harpsichord maker. He was born in Bologna, but worked in Paris. He made two harpsichords which have compasses that reflect French musical requirements rather than Italian building practice.

Ferrini, Giovanni (*fl* 1699–1755). Italian harpsichord maker. He was a pupil of Cristofori and made some instruments after Cristofori's designs, such as a *cembalo traverso*. He also constructed a harpsichord-piano (see Chapter Four, §5).

Fleischer. German family of instrument makers, active in Hamburg. Hans Christoph (1638–*c*1690) was chiefly a lute maker, but is believed to have made harpsichords as well. His son Johann Christoph (1676–*c*1724) made lutes and viols but specialized in keyboard instruments. In 1718 he invented a 'Lautenclavecin' which was a gut-strung harpsichord and a 'Theorbenflügel', a similar instrument but with an additional set of metal strings (see Chapter Four, §7). Only one harpsichord by him survives (1710; in the Staatliches Institut für Musikforschung, Berlin). A harpsichord and a clavichord by his brother Carl Conrad (1680–*c*1738) survive.

Franciolini, Leopoldo (1844–1920). Italian dealer in and forger of antique instruments. His importance lies in the fact that he was active when several of the world's largest public and private collections were being formed. Consequently, his faked and altered instruments are found in many museums and have been pictured and described in reference works with the names and dates of their purported makers.

Franciscus Patavinus (Ongaro) (*fl c*1570). Italian harpsichord maker, probably of Hungarian origin. Few of his instruments have survived, but a harpsichord in the Deutsches Museum, Munich, shows the finest possible workmanship.

Friederici [Friedrichs]. German family of keyboard instrument makers (including organs). Christian Ernst (1709–80) made an early pyramid piano, in cooperation with his brother Christian Gottfried (1714–77); several examples survive. Their sons and grandsons continued the family tradition. C. P. E. Bach praised C. E. Friederici's clavichords and preferred the absence of 4′ strings in the bass, then common in instruments by Fritz and Hass.

Fritz, Barthold (1697–1766). German instrument maker. He built organs, positives, harpsichords, clavichords, pianos and mechanical instruments of various kinds. His reputation was based on the fine quality of his clavichords.

Gaveau. French firm of piano and harpsichord makers. Joseph Gaveau founded the firm in Paris in 1847 to make pianos. Arnold Dolmetsch introduced spinet and clavichord production during his employment there (1911–14).

Gheerdinck, Artus (*fl* 1564–1624). Flemish maker of harpsichords, active in Amsterdam. There is a virginal of 1605 by him in the Germanisches Nationalmuseum, Nuremberg. Its compass is *C/E* to *c'''*.

Giusti, Giovanni Battista (*fl* 1676–93). Italian harpsichord maker, active in Lucca. Several of his harpsichords have survived; all show good workmanship.

Goble, Robert (John) (*b* 1903). English maker of harpsichords and recorders. He learnt his craft with Arnold Dolmetsch, eventually establishing his own workshop in Haslemere, before moving to Oxford in 1947. Until 1971 his instruments were built in the modern tradition, after Dolmetsch, then followed more historical lines. His son Andrea (*b* 1931) joined the firm in 1947 and Andrea's son Anthony (*b* 1957) joined in 1975.

Goermans. A family of instrument makers working in Paris, of Flemish origin. Five harpsichords made by Johannes (Jean) Goermans (1703–77) are known. His business was continued by his son Jacques (*c*1740–1789). Three of Jacques' harpsichords are known and it is recorded that in 1781 he designed a harpsichord with 21 keys to the octave.

Goff, Thomas (Robert Charles) (1898–1975). English instrument maker. He studied the piano with Irene Scharrer, read history at Christ Church, Oxford, and was called to the bar. With J. C. Cobby, a cabinet maker, he produced harpsichords and clavichords of excellent workmanship from designs by Herbert Lambert. The harpsichords were modern in construction, with an iron frame.

Gough, Hugh (Percival Henry) (*b* 1916). English maker of clavichords, harpsichords and lutes. He took a degree in economics at University College, London (1937), after he had made his first clavichord (1935) and had lessons with Arnold Dolmetsch. After war service he made several types of instruments, including a *cembal d'amour* and pianos based on late 18th-century Viennese models. In 1959 he moved his workshop to New York, where he was also active in promoting concerts of early music and dealing in antique instruments.

Goujon, Jean Claude (*fl* 1743–58). French harpsichord maker. He worked in Paris, making harpsichords and spinets and also producing *ravalements* of Ruckers harpsichords. One instrument made by Goujon was once thought to have been by Hans Ruckers: doubtless because Goujon would have been able to command a higher price for a rebuilt Ruckers instrument than for one of his own.

Gräbner. German family of instrument makers, active in Dresden in the 17th and 18th centuries. Johann Heinrich (1665–1739) was a court organ builder and also made clavichords and harpsichords; a harpsichord of 1722, with the range *E'* to *e'''*, is in Prague. His son Johann Heinrich (*c*1700–*c*1770) also became court organ maker; his 1739 harpsichord is in Schloss Pillnitz, near Dresden. Karl August (1749–*c*1796) was the third son of the younger Johann Heinrich. After his father's death he left his brothers and set up his own firm, which produced keyboard instruments of high quality.

Grauwels, Johannes [Hans] (*fl c*1580). Flemish harpsichord builder. He is an important representative of 16th-century Flemish instrument making. One of his virginals has survived; it is tonally different from the Ruckers design since the keyboard is centred and thereby avoids the tonal extremes of the *spinett* and muselar. Lodewijck Grauwels was probably a son of Johannes.

Grimaldi, Carlo (*fl* 1697). Italian instrument maker. Two extremely long harpsichords by him are known; the example in the Germanisches Nationalmuseum, Nuremberg, has an elaborately gilded outer case.

Guarracino, Honofrio (*fl* 1661–92). Italian virginal maker, active in Naples. He made several rectangular virginals; they show careful workmanship and are not in the northern Italian tradition.

Hagaerts. Flemish family of harpsichord makers. Cornelis (*d* 1642) had a son, Simon (baptized 1613), who also became a harpsichord maker and took one of Joannes Couchet's sons as an apprentice. Four instruments are known to have survived: two virginals by Cornelis (1636 and 1641) and two harpsichords by Simon (1632, in the Paris Conservatoire; the other, undated, is in the Brussels Museum of Musical Instruments).

Harris, Baker (*fl* 1740–80). English harpsichord maker. A large number of spinets by him are known, but only two harpsichords survive. Longman & Broderip acted as agents, as probably did others.

Harris, John (*fl c*1730–69). English spinet and harpsichord maker. He was the son of Joseph Harris, also an instrument maker. In 1730 he was granted a payment for 'a new (!) invented harpsichord', whose description implies a type of instrument with only unison stringing but fitted with some octave coupler device. In 1768 he emigrated to the USA and continued business in Boston.

Hasard, John. English harpsichord maker. His name is on a harpsichord of 1622, but it is difficult to read and has previously been given as 'Haward' (see Mactaggart, 1987).

Hass. German family of wind and string instrument makers, active in Hamburg. Its members include Hieronymus Albrecht (baptized 1689; *d* between 1746 and 1761) and his son Johann Adolf (*d* by 1776). Dietrich Christoph Hass may have been related to them. Several Hass harpsichords show an attempt to develop the potential of the instrument: a 1723 harpsichord (in Copenhagen) has four stops (3 × 8′, 1 × 4′; *F′* to *c″″*) and a sliding lower manual for coupling; several have a 16′ register and even a 2′ for the lower half of the compass. The largest instrument made before the 20th century has three keyboards, five sets of strings (16′, 2 × 8′, 1 × 4′, 1 × 2′), six rows of jacks (including a lute stop), harp for the 16′ row and coupling devices. J. A. Hass is best known for his clavichords which are usually unfretted, *F′* to *f‴*, with 4′ strings in the bass. (See also Chapter One, §4(iv) and Chapter Three, §4.)

Haward. English family of spinet and harpsichord makers, of whom two, Charles and Thomas, were active in London in the 17th century. The 1683 Charles Haward harpsichord has some characteristics in common with Italian instruments. Charles also made bentside spinets of which 11 have survived. The Hawards' work is an important record of instrument making in England in the 17th century. (For John Haward, *see* HASARD, JOHN.)

Haxby, Thomas (baptized 1729; *d* 1796). English instrument maker, active in York. He made harpsichords, spinets, pianos, organs, citterns and violins, and published music. His annual production of square pianos was 24 instruments from 1787 and reached 36 in 1790. He was a finer craftsman than many of his contemporaries in London.

Hemsch, Jean Henri (*fl* 1720–69). French harpsichord maker, of German origin. He was born at Castenholz, near Cologne, and went to Paris about 1720. He worked in the rue Quincompoix where 14 harpsichords were under construction at the time of his death. Five dated harpsichords and an undated one have survived. His brother Guillaume (*fl* 1748–76) apparently worked with him at the same address.

Herz, Eric (*b* 1919). American keyboard instrument maker, of German origin. In 1938 he moved to Israel where he trained as a cabinet maker and graduated from the Jerusalem conservatory of music. He worked as a musician before emigrating to the USA in 1952. From then until 1954 he was associated with Hubbard and Dowd before setting up his own workshop. He has produced about 450 harpsichords and clavichords, most of which are north-German in style. Herz has gradually moved towards historical models in his designs, while also striving for reliability and tuning stability.

Hieronymus Bononiensis. Italian harpsichord maker working in Rome in 1521. An instrument of this date, now at the Victoria and

Albert Museum, London, is the second oldest surviving dated harpsichord (*see also* VINCENTIUS and Chapter One, §2(i)).

Hitchcock. English family of spinet and harpsichord makers. They include Thomas Hitchcock (*d* before 1700) and his son Thomas (*c*1685–after 1733). (There may also have been another Thomas.) A harpsichord of *c*1725 (see fig.16) is one of the oldest surviving English two-manual harpsichords. Judging by extant instruments, the Hitchcocks concentrated on spinet making.

Hodsdon, (Wilfred) Alec (*b* 1900). English instrument maker. He established a workshop at Lavenham, Suffolk, for the restoration and production of early types of instruments, including string keyboard instruments, lutes, cornetts, regals and positive organs. His early harpsichords were highly complex, but later instruments followed simpler and more historical lines.

Hollister. Irish family of organists and keyboard instrument makers, active throughout the 18th century. Its members include Robert, Thomas (*fl* 1695–1720), Philip (*d* 1760), William (*d* 1802) and Frederick (1761–after 1802).

Hubbard, Frank (Twombly) (1920–76). American harpsichord maker. He studied English literature at Harvard, but transferred his interests to musical instruments after the success of a clavichord built with William Dowd. He went to England to learn harpsichord making, briefly at the Dolmetsch workshop and then with Hugh Gough. He also studied early instruments in European collections. In 1949 he established with Dowd a workshop to build harpsichords on historical principles rather than in the modern fashion practised by virtually all makers at that time. The partnership continued until 1958. Meanwhile Hubbard had undertaken the research that led to the publication in 1965 of his authoritative *Three Centuries of Harpsichord Making.* This book provided the most systematic and comprehensive study of harpsichord making yet published and has been a foundation for much subsequent work by other authors. In 1967 he was asked to establish a restoration workshop for the Musée Instrumental at the Paris Conservatoire. Restoration was also undertaken in his Boston workshop and many instrument makers were trained there. Hubbard's own production of finished instruments was necessarily limited, but he developed a kit based on the 1769 Taskin harpsichord, of which about 1000 were produced by 1975 (see fig.24). He also restored a number of early violins to their pre-19th-century state and made bows of a pre-Tourte type.

Hubert, Christian Gottlob (1714–93). German maker of clavichords, organs, harpsichords and pianos, of Polish origin. He became court instrument maker at Ansbach. He was best known for his clavichords, many of which were fretted until as late as 1787. His pianos were exported to France, England and the Netherlands (see Strack, 1979).

Hyman, William (1931–74). American harpsichord maker. He learnt to play the organ and was attracted to the harpsichord through a recital by Wanda Landowska. Basically self-taught, he became an accomplished woodworker, decorator and gilder and from 1963 until his death he made harpsichords in Hoboken, New Jersey. While serving with the US armed forces in Germany, he studied old harpsichords in museum collections and assimilated the old craft traditions. With his wife Doris and (from 1972) a few apprentices he made about 25 harpsichords, which have been acclaimed for their full tone and light touch. In 1971 Hyman became adviser to David Way, then head of Zuckermann harpsichords.

Jacquet. French family of instrument makers and musicians, of whom Jehan (*d* after 1658) and Claude (baptized 1605; *d* before 1675) were harpsichord makers. A two-manual harpsichord of 1652 by Claude Jacquet survives.

Karest, Joes (*c*1500–*c*1559). Flemish harpsichord maker, of German origin. He became a citizen of Antwerp in 1516–17 and was received into the Guild of St Luke in 1523. He headed the harpsichord makers who in 1557 petitioned the guild for admission as instrument makers rather than as painters. He was not only one of the founder members of the profession within the guild, but also important in the development of Flemish instrument making. Two virginals have survived (1548, see fig.26; and 1550) which permit comparison with Ruckers and other Flemish makers.

Keene, Stephen (*c*1640–*c*1719). English spinet maker working in London. A large number of his spinets and two rectangular virginals, which are typical of the 17th-century English style, have survived.

Kirckman. English family of harpsichord and piano makers, of Alsatian origin. Jacob Kirckman (1710–92) went to England in the early 1730s and worked for Hermann Tabel, becoming a British citizen in 1755. In about 1770 he went into partnership with his nephew Abraham (1737–94). Abraham took into partnership his son Joseph, who in turn took his own son Joseph (*c*1790–1877) into the business. The firm was absorbed by Collard in 1896. Kirckman's harpsichords, in common with some other English instruments, are distinctive for their veneered finish with cross-banding. There were three main types: singles of 2 × 8′; singles of 2 × 8′, 1 × 4′; and doubles of 2 × 8′, 1 × 4′, with a lute stop. Usually there was a buff stop and the compass was normally F' to f'''. Kirckman and Shudi were the most important English makers in the second half of the 18th century and together appear to have had a near monopoly on supplying harpsichords. Over a hundred Kirckman instruments have survived. (See Chapter One, §4(ii).)

Lambert, Herbert. English clavichord maker, active in Bath in the early 20th century; his designs were followed by Thomas Goff.

Longman & Broderip. English firm of music publishers and instrument makers. It was founded in or before 1767 by James Longman and others. The harpsichords were of the standard Kirckman–Shudi type (one was acquired by George Washington and is at Mount Vernon). The firm offered pianos and other instruments, but became bankrupt in 1798; the partners, with other associates, continued in instrument making.

Loosemore, John (1613 or 1614–1681). English organ builder and virginal maker. A virginal built by him in 1655 is in the Victoria and Albert Museum, London.

Maendler–Schramm. German firm of harpsichord and piano makers. Karl Maendler (1872–1958) began as a piano maker, and through marriage became head of the piano firm of M. J. Schramm in Munich. Maendler's first harpsichord was produced in 1907 and he continued the production of harpsichords and clavichords until he became blind in 1956. He was one of the first to build the heavily-framed so-called 'Bach disposition' harpsichord for mass production.

Mahoon, Joseph (*fl* 1729–71). Spinet and harpsichord maker who worked in Golden Square, London. He was appointed 'harpsichord maker to His Majesty' in 1729. Nearly all his surviving instruments are spinets. Harpsichords of 1738 and 1742 survive: their specification is F', G' to f''', with $2 \times 8'$, lute stop, and $1 \times 4'$.

Marius, Jean (*fl* 1700–16). French harpsichord maker and inventor. He became known for his invention in 1700 of the *clavecin brisé*, a travelling harpsichord, made in three hinged sections, that could be folded up into a rectangular box-shape. Four examples survive. In 1716 he submitted four designs to the Académie des Sciences for *clavecins à maillets*, i.e. harpsichords with hammers, but these designs were never realized.

Mayer, Johann (*fl* 1619). German maker of organs and other keyboard instruments, active in Salzburg. A harpsichord of 1619 with a $2 \times 8'$ register, kept in Salzburg, is one of the very few 17th-century harpsichords from the German-speaking area.

Mercator [Krämer], Sir **Michael** (1491–1544). Dutch or German harpsichord maker. He was a maker of virginals to Floris, Count of Egmont. In 1526 he was in England and was included in a list of the musical establishment of Henry VIII. Between 1529 and 1532 he made 'virginals' for both Henry VIII and Cardinal Wolsey.

Merlin, John Joseph (1735–1803). Flemish instrument maker and inventor. He went to England in 1760 and in 1774 patented a 'compound harpsichord' which included a piano action (see Chapter Four, §5). He also made pianos, besides various inventions including an invalid chair (which is still used today). Charles Burney ordered a six-octave (*C′* to *c′′′′*) piano from him in 1777.

Mietke, Michael (*d* 1719). German instrument maker at Charlottenburg. In 1719 J. S. Bach went from Cöthen to Berlin to buy a large harpsichord with two manuals from him. Two instruments in Schloss Charlottenburg, Berlin, have been attributed to him.

Müller, Hans (*fl* 1537–43). German harpsichord maker, active in Leipzig. He made the oldest surviving German harpsichord (1537), which is at octave pitch and has a transposing keyboard (see Chapter One, §2(ii) and fig.6).

Neupert. German firm of piano and harpsichord makers. It was founded by Johann Christoph Neupert (1848–1921) in 1868 as piano builders. Although harpsichords, clavichords and fortepianos were not produced before 1907–8, it had begun to collect historical instruments in 1895. This collection was donated to the Germanisches Nationalmuseum, Nuremberg, in 1968. Hanns Neupert (1902–80) joined the firm as technical director in 1928 after an apprenticeship in piano making and studies in musicology and physics. He wrote a number of works dealing with historical string keyboard instruments. The firm's production has been typical of the pre-1939 modern German school, with heavy piano-type framing, open bottoms and a 16′ register in the larger instruments. Clavichords and fortepianos were modelled more closely on historical styles, and since 1970 'copies' of old instruments have been produced. Wolf Dieter Neupert (*b* 1937) studied physics, then trained as a recording engineer. He worked at the Technical University of Berlin before joining the family firm (1973), of which he has been head since 1975.

Pesaro, Domenico da. *See* DOMINICUS PISAURENSIS.

Player, John (*c*1634–*c*1706). English spinet maker. He was apprenticed to Gabriel Townsend and made several spinets. Thomas Day recorded in 1712 having seen a harpsichord by Player with 'split or quarter notes'. The instrument has not survived.

Pohlmann, Johannes (*fl* 1767–93). English harpsichord and piano maker, of German origin. He was possibly the second piano maker in London after Zumpe. No harpsichords by him survive and his earliest known instrument is a square piano dated 1767. His pianos usually had the range *F′* to *f′′′* and had two hand stops to raise the dampers in the treble and bass.

Richard, Michael (*fl* 1659–93). French harpsichord maker. There are several Parisian harpsichord makers with this name, but it is not clear how, or if, they are related. The most important was the Michael Richard who established his workshop at the Temple in 1659. He made harpsichords, spinets and *ottavini* (instruments that play at 4′ pitch; see

Chapter Two, §2). A Jean Richard was also active in Paris, as a surviving harpsichord testifies.

Ridolfi, Giacomo (*fl* 1650–82). Italian harpsichord maker. He described himself in an inscription on one of his harpsichords as a pupil of Zenti.

Rossi. Italian family of instrument makers active in Milan. Nine polygonal virginals made by Annibale Rossi (*fl* 1542–77) have survived; an exceptionally ornate instrument (now in the Victoria and Albert Museum, London) was made for the Trivulzio family. His son Ferrante continued the business.

Rother [Rowther]**, Henry William** (*fl c*1762–82). Irish organ and harpsichord builder. He made a clavicytherium (1774) which had a 2 × 8′ disposition; it is kept in the National Museum of Ireland, as is a square piano by him.

Rubio, David (Joseph) (*b* 1934). English maker of viols, violins, lutes, guitars and harpsichords. He took a medical degree at Trinity College Dublin but studied guitar playing in Spain. In the USA he became interested in instrument making and built guitars and lutes in New York. In 1967 he returned to England and began to make harpsichords. These were mostly double-manual instruments in the 18th-century French style; he later added viols and violins to his output.

Ruckers [Ruckaert, Ruckaerts, Rucqueer, Rueckers, Ruekaerts, Ruijkers, Rukkers, Rycardt]. Flemish family of harpsichord and virginal makers. In the 16th, 17th and 18th centuries their instruments influenced other makers throughout Europe, and during the 20th-century revival of harpsichord making their sound has been highly regarded and emulated. Hans Ruckers (*b c*1540–*c*1550; *d* 1598) was married in 1575 and two of his sons followed him as harpsichord makers. In 1579 he became a member of the Guild of St Luke in Antwerp. On his death his son Joannes (Hans, Jan) (baptized 1578; *d* 1642) became a partner in the business with his brother Andreas (Andries) (baptized 1579; *d* between 1651 and 1653), but in 1608 bought out his brother to become sole owner. It is not known where Andreas moved after he sold his share in the workshop, but he continued making instruments and probably taught his son, also named Andreas (Andries) (baptized 1607; *d c*1655). With the exception of one six-sided virginal made by Hans in 1591, all the Ruckers virginals are rectangular. There were six different sizes, each size tuned to a different pitch. The larger virginals were of two types: spinetten and muselars. A double instrument combined an 8′ virginal with a 4′ one (known as a 'mother and child'); the 4′ instrument was kept in a drawer in the larger virginal and could be taken out and played, or placed on top of the 8′ virginal and coupled with it so that when the 8′ instrument was played the 4′ virginal sounded as well. Ruckers harpsichords were basically of two sizes: singles, with 1 × 8′, 1 × 4′ strings and the compass *C/E* to *c‴*; and doubles, with a 'transposing' keyboard, 1 × 8′, 1 × 4′, with the lower-manual range *C/E* to *f‴*, and the upper-manual *C/E* to *c‴* (no coupling between the keyboards was possible). Although these two sizes accommodated most of the Ruckers's clients, some harpsichords with chromatic basses were made for export to England (see Chapter One, §3(i)).

Rutkowski & Robinette. American firm of harpsichord makers in Stony Creek, Conn. Frank Rutkowski (*b* 1932) served an apprenticeship (1953–6) with John Challis and in 1957 started his business with restorations for the Morris Steinert Collection at Yale University. Robert Robinette (*b* 1929) studied music; his partnership with Rutkowski began in 1961. The firm has produced about 60 harpsichords; most are modelled after those of Challis, but those built after 1977 follow traditional building practice more closely.

Schmahl. German family of organ builders and keyboard instrument makers, active in Ulm, Heilbronn and Regensburg. Friedrich Schmahl (*fl* 1692) was a maker of clavichords and other instruments. Georg Friedrich (1748–1827) worked in Ulm and made organs and other instruments. Carl Friedrich (*fl* 1790) was active in Regensburg as an instrument maker. Christoph Friedrich (1739–1814) made keyboard instruments. He married the eldest daughter of Franz Jakob Späth, who invented the *Tangentenflügel* (see Chapter Four, §9), and later became a partner in Späth's business.

Schütze, Rainer (*b* 1925). German harpsichord maker. He studied architecture and industrial design and also learnt harpsichord making in the workshop of Walter Merzdorf in Grötzingen. He established his own workshop in Heidelberg in 1954 but remained active until 1959 as an industrial designer. Schütze (together with Martin Skowroneck and Klaus Ahrend) has been a leading German exponent of harpsichord making on historical principles. He has, nevertheless, applied his industrial design training to the construction of actions, for which he has taken out patents. He has contributed papers on instrument making to several conferences.

Shortridge, John (*b* 1930). American instrument maker. He studied at Indiana University (BA 1952, MMus 1960) and became a curator of musical instruments at the Smithsonian Institution (1958–61). From 1960 he was active as a harpsichord maker in Washington, DC. Together with his wife, he has produced about 70 instruments, all of which are meticulous copies of antique originals. His booklet *Italian Harpsichord Building in the 16th and 17th Centuries* (1960) was a seminal contribution to the history of Italian string keyboard instruments.

Shudi [Schudi, Tschudi, Tshudi], **Burkat** [Burkhardt] (1702–73). English harpsichord maker of Swiss birth. He went to England in 1718 and worked at first for Hermann Tabel in London, but by 1729 had set up on his own account. In 1761 John Broadwood started working for him and in 1769 married his daughter Barbara and became a partner. Shudi's 'Venetian swell' (see Appendix One, 'Swell') was patented in 1769. Shudi retired to a house in Charlotte Street in 1771 and on his death his partnership was taken by his son Burkat (*c*1738–1803); ownership of the firm passed to Broadwood after the death of the younger Burkat. Shudi was one of the most important English harpsichord makers of his period, the tone of his instruments being preferred by Charles Burney to Kirckman's. His clients included royalty and the nobility as well as musicians such as Haydn. Three basic models were produced: a single harpsichord 2 × 8′; a single 2 × 8′, 1 × 4′; and a double 2 × 8′, 1 × 4′ and lute stop. Usually there was a buff stop, sometimes operated by a pedal. The 'machine stop' (see Appendix One) was perhaps invented for Frederick the Great's instruments. The general style of the instruments was developed from Flemish models.

Silbermann. German family of instrument makers. Gottfried Silbermann (1683–1753), one of the most renowned German organ makers, also constructed clavichords, spinets and harpsichords as well as an early fortepiano with a Cristofori action. C. P. E. Bach had a Silbermann clavichord which he greatly prized. One harpsichord in Berlin which has been attributed to Gottfried was probably made by his nephew Johann Heinrich (1722–99), who worked in Strasbourg. Apart from this one harpsichord, the other surviving instruments from Johann Heinrich's workshop are spinets and clavichords.

Skowroneck, (Franz Hermann) Martin (*b* 1926). German instrument maker. He qualified as a teacher of flute and recorder in 1950. He made recorders for himself and was soon commissioned to make instruments for other players. He made his first harpsichord in 1953. Although without formal training in this field, his study of historical instruments and his abilities have led to his being regarded as one of the foremost makers. He has produced instruments based on most traditions of harpsichord making. His output has not been large, but has become well known through recordings. He has restored several important antique instruments and he has been an intelligent and provocative contributor to the discussion of restoration and the practice of instrument building.

Sodi, Vincenzio (*fl* 1779–92). Italian harpsichord maker, active in Florence. The last surviving dated Italian harpsichord was made by him in 1792.

Southwell, William (1756–1842). Irish maker of pianos, harpsichords and harps. He was apprenticed to Ferdinand Weber in Dublin and opened a shop himself in 1782. He devoted himself mostly to the piano, for which he took out several patents.

Stein, Johann (Georg) Andreas (1728–92). German instrument maker. He is principally known as one of the pioneers of the fortepiano, but he also built organs and was an organist in Augsburg, and, like most organ makers of this period, made harpsichords. The only harpsichords known are combination instruments, with a fortepiano at the other end of a rectangular case (see Chapter Four, §5).

Tabel [Table]**, Hermann** (*d* before 8 May 1738). English instrument maker, of Flemish origin. He may have been apprenticed to the Couchets, but about 1700 he settled in London. Both Shudi and Kirckman worked for him and Kirckman married his widow. He enjoyed a considerable reputation among his contemporaries but only one (altered) harpsichord (1721) has survived (see fig. 15 and Chapter One, §3(iii)).

Taskin, Pascal (Joseph) (1723–93). French harpsichord maker. He became a senior employee in the Blanchet workshop and on the death of the younger François Etienne Blanchet in 1766 was admitted into the guild as a master; shortly thereafter he married Blanchet's widow. He built on the success of the Blanchet workshop and his own instruments were even more carefully made than those of his former employer. In the late 1760s a *peau de buffle* register was added to the standard French three-register disposition and in 1768 Taskin perfected a system of *genouillères* (knee-levers) for changing registers. He continued the Blanchet practice of rebuilding and enlarging Ruckers harpsichords. From the late 1770s his workshop was increasingly occupied with the building of grand pianos; these were well made but of primitive design, the action being without escapement. Taskin's name is today well known because his 1769 harpsichord (Russell Collection, Edinburgh; see figs. 17 and 24) has been used as a model by many makers.

Theewes [Theeuwes, Teeus]. Flemish family of harpsichord makers of the second half of the 16th century. Three members of the family are known as instrument makers: Jacob (*fl* 1533–57), Lodewijk (*fl* 1557) and Jacob's son, Lodewijk (*fl* 1560–85). The only surviving instrument was made by Jacob's son Lodewijk, who was admitted to the Antwerp Guild of St Luke in 1561 but had emigrated to England by 1568. In 1579 he made, as part of a claviorgan, what is now the earliest surviving English harpsichord. This instrument is therefore an important document of instrument making in England in the 16th century (see Koster, 1980; see also Chapter One, §2(ii).

Tibaut, Vincent (*fl* 1679–81). French provincial harpsichord maker. His importance lies in the fact that he is one of the few 17th-century French harpsichord makers whose work is well enough known for comparisons to be made with the work of makers elsewhere. Three two-manual harpsichords are known.

Todini, Michele (*c*1625–*c*1689). Italian instrument maker. He was a virtuoso on the Italian musetta, a bellows-blown chamber bagpipe, but his chief claim to distinction lies in the complicated (keyboard) instruments he made, described in Kircher (1673) and Todini (1676). Pietro Todini, a Milanese harpsichord maker, may have been his son.

Trasuntino [Trasontino, Trasuntinus]. An important group, some of them related, of Italian instrument makers. The earliest known is Alessandro (*c*1485–*c*1545), who was an organ builder and harpsichord maker. The family name of Vito (Guido, Giulio) Trasuntino (de Trasuntinis) (*c*1522–after 1606) was in fact Frassonio, so he may have adopted the name Trasuntino for business reasons. He made organs as well as string keyboard instruments. His son Claudio appears not to have become an instrument maker. The earliest surviving Venetian harpsichord (now privately owned in Italy) was made in 1530 in Alessandro's workshop. He also made arpicordi but none has survived. Vito was possibly the most famous 16th-century Italian maker: Garzoni (1585) and Fioravanti (1564) both praise his skills. He was also an organ maker and an expert who was called in to judge others' work. He is best known as the constructor of the 'Clavemusicum omnitonum' (now in the Museo Civico Medievale, Bologna; see fig.5).

Vater. German family of instrument makers. Martin Vater was an organ builder. He taught his son Christian (baptized 1679; *d* 1756), who worked with Arp Schnitger and subsequently worked on many organs; only one harpsichord by him is known. His brother Antoine (Anton) (1689–1759) went to Paris in 1715 and made harpsichords there; at least four of his instruments have survived.

Vaudry, Jean Antoine (*c*1680–1750). French harpsichord maker, active in Paris. Until recently he was known only from bibliographical references, but a harpsichord by him was discovered and is now in the Victoria and Albert Museum, London (see fig.13; see also Adlam, 1976).

Vincentius (*fl* 1515). Italian harpsichord maker. He made what is now known to be the oldest surviving harpsichord (older than that of Hieronymus Bononiensis, 1521, which is usually cited as the oldest). The instrument was made in 1515 for Pope Leo X. It has been extensively altered; its original disposition was probably 1 × 8′, with a compass of *C*/*E* to f‴ (see Wraight, in *Early Music*, 1986; see also Chapter One, §2(i) and fig.3).

Wagner, Johann Gottlob [Jean Théophile] (1741–89). German maker of keyboard instruments. He was a pupil of Silbermann before he established his workshop in Dresden. His younger brother, Christian (1754–between 1812 and 1816), joined him as a partner in 1773 and assisted him in the invention of the 'clavecin royal' in 1774. This was a square piano, usually with the compass *F*′ to *f*‴; the hammers were not covered, so that the sound resembled that of a harpsichord, and there was a 'Harfe' stop to produce a damped sound.

Weber [Webber], **Ferdinand** (1715–84). Irish keyboard instrument maker, of German origin. He learnt organ building with Johann Ernst Hähnel of Lower Meissen before moving to Dublin in 1739. His activities there included making and tuning organs, but he also made harpsichords and, from about 1772, fortepianos. An undated clavicytherium by him also survives; it has three rows of jacks and an ingenious single spring mechanism to return the jacks. Southwell was one of his apprentices; Robert Woffington may also have been his pupil.

Zell, Christian (*fl* 1722–41). German harpsichord maker, active in Hamburg. He married the widow of Carl Conrad Fleischer. Two single-manual harpsichords have survived and one two-manual instrument of 1728, now in the Museum für Kunst und Gewerbe, Hamburg. It has a compass *F′* to *d‴*, and 2 × 8′, 1 × 4′, with buff stop and coupler.

Zenti, Girolamo [Hieronimus de Zentis Viterbiensis] (*d* 1668). Italian maker of harpsichords, spinets and organs. He was the best known and most widely-travelled 17th-century Italian keyboard instrument maker working in the Swedish, English and French royal courts. The earliest signed bentside spinet bears Zenti's name and it is possible that he invented the shape of this instrument. He appears to have built several different sizes of harpsichord, but few of his instruments survive.

Zuckermann, Wolfgang Joachim (*b* 1922). American harpsichord maker, of German origin. He moved to the USA in 1938 and read psychology at Queens College, New York. His musical interests led him to study piano technology and as a piano technician he often had to repair and regulate harpsichords. In 1954 he built a simplified, single-manual harpsichord for his own use, then produced similar instruments. In 1960 he made a kit version and by the end of 1969 almost 8000 instruments had been sold. In 1969 he sold his New York enterprise and moved to England. The name has been retained for the kits produced after 1970; these were modelled more closely on historical instruments (see Zuckermann, 1969).

Zumpe, Johannes [Johann Christoph] (*fl* 1735–83). English harpsichord and piano maker, of German origin. He probably worked for the Silbermanns and was the most famous of the German makers known as the 'twelve apostles' who emigrated to London following the Seven Years War. Zumpe worked for Shudi before setting up his own workshop. It is likely that he made a few harpsichords but he is best known for square pianos which had a light action and a tone not unlike that of a large 18th-century clavichord.

APPENDIX THREE

Editions

1. Collected editions

AMI *L'arte musicale in Italia*, ed. L. Torchi (Milan, Rome, 1897–1908/*R*1968)

AMO *Archives des maîtres de l'orgue des XVIe, XVIIe et XVIIIe siècles*, ed. F. A. Guilmant and A. Pirro (Paris, 1898–1911/*R* 1972)

AMP *Antiquitates musicae in Polonia*, ed. H. Feicht, Institute of Musicology, U. of Warsaw (Graz and Warsaw, 1963–)

CEKM *Corpus of Early Keyboard Music*, ed. W. Apel and others, American Institute of Musicology (Rome, 1963–)

CMI *I classici musicali italiani*, Fondazione Eugenio Bravi (Milan, 1941–56)

CMM *Corpus mensurabilis musicae*, ed. A. Carapetyan and others, American Institute of Musicology (Rome, 1947–)

DDT *Denkmäler deutscher Tonkunst*, 1st ser., ed. R. Liliencron (1901–11), H. Kretzschmar (1912–18), H. Abert (1927) and A. Schering (1928–31), Königliche Preussischen Regierung Berufene Kommission (1892–1900), Musikgeschichtliche Kommission (1901–60) (Leipzig, 1892–31; rev. 2, ed. H. J. Moser, Wiesbaden, Graz, 1957–60)

DM *Documenta musicologica*, Internationale Gesellschaft für Musikwissenschaft, 1st ser.: *Druckschriften-Faksimiles* (Kassel and Basle, 1951–), 2nd ser.: *Handschriften-Faksimiles* (Kassel and Basle, 1955–)

DTB *Denkmäler der Tonkunst in Bayern* (= DDT, 2nd ser.), ed. A. Sandberger, Gesellschaft zur Herausgabe von Denkmälern der Tonkunst in Bayern, Leipzig (Leipzig (vols.i–xx), Augsburg (xxi–xxx), 1900–31; rev. 2/1962–)

DTÖ *Denkmäler der Tonkunst in Österreich*, ed. G. Adler (vols.i–lxxxiii) and E. Schenk (lxxxiv–), Gesellschaft zur Herausgabe, Vienna (Vienna, 1894–1904, 1919–38; Leipzig, 1905–13, 1919–23; Graz, 1966–)

EDM *Das Erbe deutscher Musik*, 1st ser.: *Reichsdenkmäler*, Institut für Deutsche Musikforschung, Berlin (1935–43), Musikgeschichtliche Kommission (1953–)

EKM *Early Keyboard Music*; vols.i–xii also known as *English Keyboard Music* (London, 1951–)
EMN *Exempla musica neerlandica*, Vereniging voor Nederlandse Muziekgeschiedenis (Amsterdam, 1964–)
MB *Musica britannica*, ed. A. Lewis and others, Royal Musical Association (London, 1951–, rev. 2/1954–)
MMBel *Monumenta musicae belgicae*, ed. J. Watelet (vols.i–vii) and R. B. Lenaerts (viii–xi), Vereniging voor Muziekgeschiedenis, Antwerp, and Seminarie voor Muziekwetenschap of Louvain U. (vols.viii–xi) (Berchem and Antwerp, 1932–51, 1960–66/*R*)
MME *Monumentos de la música española*, Instituto Español de Musicología, Consejo Superior de Investigaciones Cientificas (Madrid and Barcelona, 1941–)
MMI *Monumenti di musica italiana*, 1st ser.: *Organo ed cembalo*, ed. O. Mischiati, G. Scarpat and L. F. Tagliavini (Brescia and Kassel, 1961–)
MMN *Monumenta mūsica neerlandica*, Vereniging voor Nederlandse Muziekgeschiedenis (Amsterdam, 1959–)
MMP *Monumenta musicae in Polonia*, ed. J. M. Chomiński, Instytut Sztuki Polskiej Akademii Nauk (Warsaw, 1964–)
MSD *Musicological Studies and Documents*, ed. A. Carapetyan, American Institute of Musicology (Rome, 1951–)
NM *Nagels Musik-Archiv* (Hanover and Kassel; later Kassel, London and New York, 1927–)
PÄMw *Publikationen älterer praktischer und theoretischer Musikwerke vorzugsweise des XV. und XVI. Jahrhunderts*, ed. R. Eitner, Gesellschaft für Musikforschung (Berlin, later Leipzig, 1873–1905/*R*1967)
PM *Portugaliae musica*, ser. A, Fundação Calouste Gulbenkian (Lisbon, 1959–)
PSFM *Publications de la Société Française de Musicologie*, 1st ser.: *Monuments de la musique ancienne* (Paris, 1925–)
SMd *Schweizerische Musikdenkmäler*, Schweizerische Musikforschende Gesellschaft (Kassel and Basle, 1955–)
UVNM *Uitgave van oudere Noord-Nederlandse Meesterwerken*, Vereniging voor Nederlandse Muziekgeschiedenis (Amsterdam, 1869–); vols.i–xviii with title *Maatschappij tot bevordering der toonkunst*

2. Anthologies

Alphabetical list, excluding the article.

Altenglische Orgelmusik, ed. D. Stevens (Kassel and Basle, 1953)
Altes Spielbuch aus der Zeit um 1500: Liber Fridolini Sichery, ed. F. J. Giesbert (Mainz, 1936)

EARLY KEYBOARD

Altitalienische Versetten in allen Kirchentonarten für Orgel oder andere Tasteninstrumente, ed. M. S. Kastner (Mainz, 1957)

Anne Cromwell's Virginal Book, 1638, ed. H. Ferguson (London, 1974)

Anonymi der norddeutschen Schule: sechs Praeludien und Fugen, Organum, 4th ser., x (Leipzig, 1925/*Rc*1960)

Antologia de organistas clásicos españoles, ed. F. Pedrell (Madrid, 1908/*R*1968 as *Anthology of Classical Spanish Organists*)

Antología de organistas do século XVI, PM, ser. A, xix (1969)

Antología de organistas españoles del siglo xvii, ed. H. Anglès (Barcelona, 1965–7)

Antologia di musica antica e moderna per pianoforte, ed. G. Tagliapietra (Milan, 1931–2)

Antologia organistica italiana (*sec. XVI–XVII*), ed. S. dalla Libera (Milan, 1957)

Attaingnant, P., ed., 7 collections of anonymous pieces (Paris, 1530–31): see below, *Chansons und Tänze* [i–iv]; *Deux livres d'orgues* [v–vi]; *Keyboard Dances* [iv]; *Transcriptions of Chansons* [i–iii]; *Treize motets* [vii]

Balli antichi veneziani per cembalo [Biblioteca Nazionale Marciana, Venice, Ital.IV.1227], ed. K. Jeppesen (Copenhagen, 1962)

Buxheimer Orgelbuch, DM, 2nd ser., *Handschriften-Faksimiles*, i (1955); ed. B. A. Wallner, EDM, xxxvii–xxxix (1958–9)

Cantantibus organis: Sammlung von Orgelstücken alter Meister, ed. E. Kraus (Kassel, 1958–)

Chansons françaises pour orgue vers 1550, Le pupitre, v (Paris, 1968)

Chansons italiennes de la fin du XVIe siècle, ed. A. Wotquenne-Plattel (Leipzig, n.d.)

Chansons und Tänze: Pariser Tabulaturdrucke für Tasteninstrumente aus dem Jahre 1530 von Pierre Attaingnant, ed. E. Bernoulli (Munich, 1914)

Choralbearbeitungen aus der Tabulatur Lynar A 1, ed. H. J. Moser and T. Fedtke (Kassel and Basle, 1956)

Choralbearbeitungen und freie Orgelstücke der deutschen Sweelinck-Schule, aus den Tabulaturen Lynar B 1, B 3, B 6, und Graues Kloster Ms. 52, ed. H. J. Moser and T. Fedtke (Kassel and Basle, 1955)

Classici italiani dell'organo, ed. I. Fuser (Padua, 1955)

Clavichord Music of the Seventeenth Century, ed. T. Dart (London, 1960, rev. 2/1964)

Clement Matchett's Virginal Book (1612), EKM, ix (London, 1957)

Composizioni per organo o cembalo secoli xvi, xvii e xviii, AMI, iii (n.d.)

Cravistas portuguezes, ed. M. S. Kastner (Mainz, 1935–50)

Deux livres d'orgue parus chez Pierre Attaingnant en 1531, ed. Y. Rokseth, PSFM, i (1925)

Dublin Virginal Manuscript, ed. J. Ward (London, 1983)

Dutch Keyboard Music of the Sixteenth and Seventeenth Centuries, MMN, iii (1961)

Early English Keyboard Music, ed. H. Ferguson (London, 1971)

Early Fifteenth-century Italian Source of Keyboard Music: the Codex Faenza, Biblioteca Comunale, 117, MSD, x (1961)

Early French Keyboard Music, ed. H. Ferguson (London, 1966)

Early German Keyboard Music, ed. H. Ferguson (London, 1970)

Early Italian Keyboard Music, ed. H. Ferguson (London, 1968)

Early Scottish Keyboard Music, EKM, xv (1958)

Early Spanish Keyboard Music, ed. B. Ife and R. Truby, 3 vols. (Oxford, 1986)

Early Spanish Organ Music, ed. J. Muset (New York, 1948)

Elizabethan Virginal Music, ed. H. F. Redlich (Vienna, 1938)

Elizabeth Rogers' Virginal Book 1656, ed. C. J. Cofone (New York, 2/1982)

English Court and Country Dances of the Early Baroque from MS Drexel 5612, ed. H. Gervers, CEKM, xliv (1982)

English Pastime Music, 1630–1660, ed. M. Maas (Madison, 1974)

European Organ Music of the Sixteenth and Seventeenth Centuries, Faber Early Organ Series, ed. J. Dalton, 18 vols. (London, 1986–9) [i–iii, England; iv–vi, Spain and Portugal; vii–ix, France; x–xii, Netherlands and North Germany; xiii–xv, Southern Germany and Austria; xvi–xviii, Italy]

Faenza Codex: see *Early Fifteenth-century Italian Source*; *Keyboard Music of the Late Middle Ages*

Fitzwilliam Virginal Book, ed. J. A. Fuller Maitland and W. B. Squire (Leipzig, 1899/*R*)

Fitzwilliam Virginal Book, Twenty-four Pieces, EKM, xvi (1964)

Frottole intabulate da sonare organi, Andrea Antico, 1517, ed. C. Hogwood (Tokyo, 1984)

Frühmeister der deutschen Orgelkunst, ed H. J. Moser and F. Heitmann (Leipzig, 1930)

Harpsichord Master: [i] (1697), ed. R. Petre (Wellington and London, 1980); ii (1700) and iii (1702), ed. R. Rastall (Clarabricken, Co. Kilkenny, Ireland, 1980) [facs.]

Harpsichord Pieces from Dr. Bull's Flemish Tablature, ed. H. F. Redlich (Wilhelmshaven, 1958)

Intabolatura nova di balli (*Venice, 1551*), EKM, xxiii (1965)

Intabulatura nova (*Venedig 1551*), ed. F. Cerha (Vienna and Munich, 1975)

Italienische und süddeutsche Orgelstücke des frühen 17. Jahrhunderts, Die Orgel, 2nd ser., ix (Leipzig, 1957)

Keyboard Dances from the Earlier 16th Century, ed. D. Heartz, CEKM, viii (1965)

Keyboard Music at Castell'Arquato, CEKM, xxxvii (1975)

Keyboard Music from Polish Manuscripts, CEKM, x (1965–7)

Keyboard Music of the Fourteenth and Fifteenth Centuries, CEKM, i (1963)

Keyboard Music of the Late Middle Ages in Codex Faenza 117, CMM, lvii (1972)

Klaviertänze des 16. Jahrhunderts, ed. H. Halbig (Stuttgart, 1928)
Liber Fratrum Cruciferorum, AMO, x (1909–11)
Libro de cifra nueva . . . [ed.] Luys Venegas de Henestrosa (Alcalá . . ., 1557), MME, ii (1944)
Locheimer Liederbuch nebst der Ars Organisandi von Conrad Paumann, ed. F. W. Arnold and H. Bellermann (Leipzig, 1926/*R*)
Locheimer Liederbuch und Fundamentum Organisandi des Conrad Paumann, ed. K. Ameln (Berlin, 1925) [facs.]
Lublin Tablature [Tablatura organowa Jana z Lublina], MMP, ser. B, i (1964) [facs.]; ed. J. R. White, CEKM, vi/1–6 (1964–7)
Lüneburger Orgeltabulatur KN 208a, EDM, xxxvi, xl (1957)
Melothesia, Keyboard Suites, ed. C. Hogwood (London, 1987); see also §3 below, Locke, M.
Mulliner Book, MB, i (1951, rev. 2/1954)
Mulliner Book, Eleven Pieces, EKM, iii (1951)
Musick's Hand-maid (1663), ed. T. Dart (London, 1970); (1689), ed. T. Dart, EKM, xxviii (1969)
Musik aus früher Zeit für Klavier (1350–1650), ed. W. Apel (Mainz, 1934)
Neapolitan Keyboard Composers c. 1600, CEKM, xxiv (1967)
Nederlandse klaviermuziek: see *Dutch Keyboard Music*
Old Spanish Organ Music, ed. C. Reiss (Copenhagen, 1960)
Organ Compositions by Sweelinck and Scheidt, UVNM, iii (1871)
Organo italiano, 1567–1619, ed. G. Frotscher (Copenhagen, 1960)
Organ Tablatures of Warsaw, Musical Society I/200, AMP, xv (1968)
Orgel-Meister I, Organum, 4th ser., ii (Leipzig, 1925/*Rc*1960)
Orgel-Meister IV, Organum, 4th ser., xxi (Leipzig, 1925/*Rc*1960)
Orgelmeister der Gotik, Liber organi, viii (Mainz, 1938)
Orgeltabulatur von 1448 des Adam Ileborgh aus Stendal, ed. G. Most (Stendal, 1954) [facs.]
Parthenia, ed. E. F. Rimbault (London, 1847); ed. M. Glyn (London, 1927); ed. O. E. Deutsch (London, 1942) [facs.]; ed. K. Stone (New York, 1951); EKM, xix (1960, rev. 2/1962)
Parthenia In-violata, ed. T. Dart and R. J. Wolfe (New York, 1961) [facs.]
Pelplin Tablature, AMP, i–x (1963–70)
Priscilla Bunbury's Virginal Book, Sixteen Pieces, ed. J. L. Boston (London, 1962)
Seventeenth-century Keyboard Music in the Chigi Manuscripts of the Vatican Library, CEKM, xxxii (1968)
Seven Virginal Pieces from B.M.Add.30486, Schott's Anthology of Early Keyboard Music, iii (1951)
Silva ibérica de música para tecla de los siglos XVI, XVII y XVIII, ed. M. S. Kastner (Mainz, 1954–65)
Spanish Netherlands Keyboard Music, i–ii, ed. R. Vendome (Oxford, 1983)

Spanish Organ Masters after Antonio de Cabezón, CEKM, xiv (1971)
Spielbuch für Kleinorgel oder andere Tasten-Instrumente, ed. W. Auler (Leipzig, 1942)
Tablature of Celle 1601, CEKM, xvii (1971)
Tabulaturen des XVI. Jahrhunderts, SMd, vi–vii (1967–70)
Ten Pieces by Hugh Aston and others [British Library, London, Roy.App.58], Schott's Anthology of Early Keyboard Music, i (1951)
Tisdale's Virginal Book, EKM, xxiv (1966)
Tomkins Manuscript, Pieces from the [British Library, London, Add. 29996], Schott's Anthology of Early Keyboard Music, iv (1951)
Transcriptions of Chansons for Keyboard, ed. A. Seay, CMM, xx (1961)
Treize motets et un prélude pour orgue parus chez Pierre Attaingnant en 1531, ed. Y. Rokseth, PSFM, v (1930)
Twelve Pieces from Mulliner's Book [British Library, London, Add. 30513], Schott's Anthology of Early Keyboard Music, ii (1951)
Twenty-five Pieces for Keyed Instruments from Cosyn's Virginal Book, ed. J. A. Fuller Maitland and W. B. Squire (London, 1923)
Two Elizabethan Keyboard Duets, ed. F. Dawes (London, 1949) [by T. Tomkins and N. Carleton (the younger)]

3. Renaissance, Baroque and Classical composers

Ammerbach, E. N.: *Orgel oder Instrument Tabulaturbuch (1571/1583)*, ed. C. Jacobs (Oxford, 1984); pieces ed. in W. Merian, *Der Tanz* (see Bibliography, 'Repertory', §4)
Aston, H.: *Hornepype* and other attrib. pieces, ed. F. Dawes; see §2 above, *Ten Pieces*
Bach, C. P. E.: *Die sechs Sammlungen von Sonaten, freien Fantasien und Rondos für Kenner und Liebhaber* WQ55–9, 61, ed. C. Krebs (Leipzig, 1895, rev. 2/1953 by L. Hoffmann-Erbrecht)
—— : *Die* [6] *preussischen Sonaten für Klavier* WQ48, ed. R. Steglich, NM, vi, xv (Hanover, 1927–8/*R*)
—— : *Die* [6] *württembergischen Sonaten für Klavier* WQ49, ed. R. Steglich, NM, xxi, xxii (Hanover, 1928/*R*)
—— : *Selected Keyboard Works*, iv [6 sonatas WQ63, from *Versuch über die wahre Art das Clavier zu Spielen*], ed. H. Ferguson (London, 1983)
Bach, J. C.: Sonatas: op.5, op.12/17 (*R*1976) [facs., ed. C. Hogwood]; op.5 nos.2–6 and op.12/17 nos.2–6, ed. L. Landshoff (Leipzig, 1925–7)
Bach, J. S.: Forty-eight Preludes and Fugues [*Das wohltemperirte Klavier*], ed. F. Tovey (London, 1924) [incl. notes on each work]
—— : French Suites BWV812–17, ed. R. Jones (London, 1985)
—— : Goldberg Variations [*Keyboard Practice consisting of an Aria with thirty Variations for the Harpsichord with two Manuals*], ed. R. Kirkpatrick (New York, 1938, repr. 1956); ed. K. Gilbert (Paris, 1979)

—— : Inventions and Sinfonias BWV772–801, ed. R. Jones (London, 1984)
—— : *Klavier-Werke*, ed. H. Bischoff, 7 vols. (Leipzig, 1880–88; part repr. New York, 1946)
—— : *Neue Ausgabe sämtlicher Werke (Neue Bach-Ausgabe)*, ed. Johann-Sebastian-Bach-Institut, Göttingen, and Bach-Archiv, Leipzig, ser. I–VIII (Kassel and Basle, 1954–) [keyboard works in ser. V]
—— : Partitas BWV825–30, ed. R. Jones (London, 1981)
—— : *Werke*, ed. Bach-Gesellschaft, i–xlvii (Leipzig, 1851–99/*R*1947) [keyboard works in vols.iii, xiii/2, xiv, xxxvi, xlii, xlv]
Bach, W. F.: [8] *Fugen* F31, [12] *Polonaisen* F12; ed. W. Niemann (Leipzig, 1914)
—— : [9] *Klavierfantasien* F14–21, 23, ed. P. Schleuning (Mainz, 1972)
—— : *Twelve Polonaises*, ed. R. Jones (London, 1987)
—— : *Sämtliche* [9] *Klaviersonaten* F1–9, ed. F. Blume, NM, lxiii, xviii, clvi (1930–41)
Bariolla, O.: *Keyboard Compositions*, ed. C. W. Young, CEKM, xlvi (1986)
Battiferri, L.: *Ricercari*, ed. G. G. Butler, CEKM, xlii (1981)
Blitheman, J. W.: Pieces in the Mulliner Book; ed. D. Stevens, MB, i [nos.22, 27, 32, 49–52, 77, 91–6, 108] (1951, rev. 2/1962)
Blow, J.: *John Blow's Anthology*, *c*1700, ed. T. Dart and D. Moroney (London, 1978) [works by Blow, Froberger, Fischer and Strungk]
—— : *Six Suites*, ed. H. Ferguson (London, 1965)
Böhm, G.: *Sämtliche Werke: Klavier- und Orgelwerke* [i–ii], ed. J. Wolgast, rev. G. Wolgast (Wiesbaden, 1952) [works for clavichord and harpsichord in vol.i]
Buchner, H.: *Sämtliche Orgelwerke*, ed. J. H. Schmidt, EDM, liv–lv (1974)
—— : Pieces ed. in W. Merian, *Der Tanz* (see Bibliography, 'Repertory', §4)
Bull, J.: *Keyboard Music I*, ed. J. Steele, F. Cameron and T. Dart, MB, xiv (1960, rev. 2/1967); *II*, ed. T. Dart, MB, xix (1963, rev. 2/1970)
Buus, J.: *Orgelwerke*, i–ii, ed. T. D. Schlee (Vienna, 1980–83)
Buxtehude, D.: *Klavervaerker*, ed. E. Bangert (Copenhagen, 1942, 2/1944)
Byrd, W.: *Keyboard Music I, II*, ed. A. Brown, MB, xxvii–xxviii (1969–71, rev. 2/1976)
—— : *My Ladye Nevells Book*, ed. H. Andrews (London, 1926/*R*)
Cabezón, A. de: Collected Works, ed. C. G. Jacobs, *Gesamtausgaben*, iv (New York, 1967–76)
—— : *Obras de música para tecla, arpa y vihuela*, ed. H. Anglès, MME, xxvii–xxix (1968)
Cavaccio, G.: *Sudori musicali* (1626), ed. J. Evan Kreider, CEKM, xliii (1984)

Cavazzoni, G.: *Orgelwerke I, II*, ed. O. Mischiati (Mainz, 1959–61)
Cavazzoni, M. A.: *Recerchari, motetti, canzoni* (1523), ed. in K. Jeppesen, *Die italienische Orgelmusik am Anfang des Cinquecento* (Copenhagen, 1943; rev., enlarged, 2/1960)
Chambonnières, J. C. de: *Oeuvres complètes*, ed. P. Brunold and A. Tessier (Paris, 1925/*R*1967)
—— : *Pièces de clavessin* (Paris, 1670/*R*1967); ed. T. Dart as *Les deux livres de clavecin* (Monaco, 1969)
Clarke, J.: Pieces in *The Harpsichord Master*, iii (1702), ed. R. Rastall (Clarabricken, Co. Kilkenny, Ireland, 1980) [facs.]
Clérambault, L.-N.: *Pièces de clavecin*, ed. P. Brunold, rev. T. Dart (Monaco, 1964)
Couperin, F.: *L'art de toucher le clavecin*, ed. M. Halford (New York, 1974) [Fr. and Eng. text]
—— : *Pièces de clavecin*, ed. K. Gilbert, Le pupitre, xxi–xxiv (Paris, 1969–72)
Couperin, L.: *Pièces de clavecin*, ed. P. Brunold, rev. D. Moroney (Monaco, 1985)
Croft, W.: *Complete Harpsichord Works*, ed. H. Ferguson and C. Hogwood, 2 vols. (London, 2/1981–2)
Dagincour, F.: *Pièces de clavecin*, ed. H. Ferguson (Paris, 1969)
Dandrieu, J.-F.: *Trois livres de clavecin*, ed. P. Aubert and B. Francois-Sappey (Paris, 1973)
D'Anglebert, J.-H.: *Pièces de clavecin*, ed. K. Gilbert, Le pupitre, liv (Paris, 1975)
Daquin, L.-C.: *Noëls*, ed. C. Hogwood (London, 1984)
—— : *Pièces de clavecin*, ed. C. Hogwood (London, 1982)
Della Ciaia, A. B.: *Tre sonate*, ed. G. Buonamici (Florence, 1912)
Durante, F.: *Studii, divertimenti e toccate*, ed. B. Paumgartner (Kassel, 1949)
Erbach, C.: *Collected Keyboard Compositions*, ed. C. G. Rayner, CEKM, xxxvi/1–4 (1971–7)
Facoli, M.: *Intavolatura di balli*, ed. W. Apel in *Collected Works*, CEKM, ii (1963)
Farnaby, G.: 54 pieces (1 doubtful), ed. R. Marlow, MB, xxiv (1965)
Fischer, J. C. [K.] F.: *Sämtliche Werke für Klavier und Orgel*, ed. E. von Werra (Leipzig, 1901/*R*1965)
Frescobaldi, G.: *Il Primo libro di toccate d'intavolatura di cembalo e organo, Opere complete*, ii, ed. E. Darbellay (Milan, 1977) [with facs.]
—— : *Il Secondo libro di toccate d'intavolatura di cembalo e organo, Opere complete*, iii, ed. E. Darbellay (Milan, 1980) [with facs.]
—— : *Keyboard Compositions Preserved in Manuscripts*, ed. W. R. Shindle, CEKM, xxx/1–3 (1968)
—— : *Orgel- und Klavierwerke*, ed. P. Pidoux, 5 vols. (Kassel, 1949–54)
Froberger, J. J.: *Oeuvres complètes pour clavecin*, ed. H. Schott (Paris, 1979–)

—— : *Orgel- und Klavierwerke*, ed. G. Adler, DTÖ, viii, Jg.iv/1; xiii, Jg.vi/2; xxi, Jg.x/2 (1897–1903/*R*1959)

Gabrieli, A.: *Orgel- und Klavierwerke*, ed. P. Pidoux, 5 vols. (Kassel, 1941–53)

Gabrieli, G.: *Composizioni per organo*, ed. S. Dalla Libera, 3 vols. (Milan, 1957–9)

Gibbons, O.: *Keyboard Music*, ed. G. Hendrie, MB, xx (1962)

Greene, M.: *A Collection of Lessons for the Harpsichord* (1750), ed. D. Moroney (London, 1977) [facs.]

Handel, G.F.: *Klavierwerke*, ed. R. Steglich, P. Northway and T. Best, Hallische Händel-Ausgabe, iv/1, iv/5, iv/6, iv/17 (Kassel, 1955–75)

—— : *Twenty Overtures in Authentic Keyboard Arrangements*, ed. T. Best, 3 vols. (London, 1986)

Hassler, H. L.: *Toccatas*, ed. S. Stribos, CEKM, xlv (1985)

Haydn, J.: *Werke für das Laufwerk (Flötenuhr)*, ed. E. F. Schmid (Hanover, 1931; rev., enlarged, 2/1954)

—— : *Klavierstücke*, ed. F. Eibner and G. Jarecki (Vienna, 1975)

—— : *Sämtliche Klaviersonaten*, ed. C. Landon, 4 vols. (Vienna, 1964–6)

—— : *Werke*, xviii, ed. J. Haydn-Institut, Cologne (Munich, 1960–)

Ileborgh, A.: Ed. in *Keyboard Music of the Fourteenth and Fifteenth Centuries*, CEKM, i (1963); see also §2 above, *Orgeltabulatur von 1448*

Jan z Lublina [John of Lublin, Johannes de Lublin]: see §2 above, *Lublin Tablature*

Kotter, H.: Ed. in W. Merian, *Der Tanz* (see Bibliography, 'Repertory', §4)

Kuhnau, J.: *Klavierwerke*, ed. C. Päsler, DDT, iv (1901, rev. 2/1958)

—— : *Six Biblical Sonatas*, ed. K. Stone (New York, 1953)

Lebègue, N.-A.: *Oeuvres de clavecin*, ed. N. Dufourcq (Monaco, 1956)

Le Roux, G.: *Pieces for Harpsichord* (1705), ed. A. Fuller (New York, 1959)

Locke, M.: *Keyboard Suites*, ed. T. Dart (London, 1959, rev. 2/1964)

—— : *Melothesia*, ed. C. Hogwood (Oxford, 1987); see also §2 above, *Melothesia*

—— : *Organ Voluntaries*, ed. T. Dart (London, 1957)

Loeffelholz von Colberg, C.: 22 pieces ed. in W. Merian, *Der Tanz* (see Bibliography, 'Repertory', §4)

Lublin, Johannes de: see §2 above, *Lublin Tablature*

Marcello, B.: [12] *Sonates pour le clavecin*, ed. L. Sgrizzi and L. Bianconi, Le pupitre, xxviii (Paris, 1971)

Marchand, L.: *Pièces de clavecin*, ed. T. Dart, rev. D. Moroney (Monaco and Paris, 1987)

Martini, G. B.: *Dodici sonate per cembalo ov organo* (1742), ed. M. Vitali (Milan, *c*1927)

Mayone, A.: *Diversi capricci per sonare* (i, 1603; ii, 1609), ed. C. Stembridge (Padua, 1981–4)
Merulo, C.: *Canzoni d'intavolatura d'organo, 1592*, ed. P. Pidoux (Kassel and Basle, 1941/*R*1954)
—— : *Toccate per organo*, i–iii, ed. S. Dalla Libera (Milan, 1959/*R*)
Muffat, Georg: *Apparatus musico-organisticus*, ed. R. Walter (Altötting, 1957)
Nörmiger, A.: 69 pieces ed. in W. Merian, *Der Tanz* (see Bibliography, 'Repertory', §4)
Paix, J.: 20 pieces ed. in W. Merian, *Der Tanz* (see Bibliography, 'Repertory', §4)
Pasquini, B.: *Collected Works for Keyboard*, ed. M. B. Haynes, CEKM, v/1–7 (1964–8)
Paumann, C.: *Fundamentum organisandi*, ed. W. Apel, CEKM, i (1963)
Picchi, G.: *Complete Keyboard Works*, ed. H. Ferguson (Tokyo, 1979)
Poglietti, A.: *Rossignolo* suite and other pieces, ed. in DTÖ, xxvii, Jg.xiii/2 (1906/*R*)
Purcell, H.: *Eight Suites*; *Miscellaneous Keyboard Pieces*, ed. H. Ferguson (London, 2/1968)
Radino, G. M.: *Il primo libro d'intavolatura di balli d'arpicordo*, ed. S. Ellingworth, CEKM, xxxiii (1968)
Rameau, J.-P.: *Pièces de clavecin*, ed. K. Gilbert (Paris, 1979)
Roseingrave, T.: *Compositions for Organ and Harpsichord*, ed. D. Stevens (University Park, Penn., 1964) [includes 4 suites]
—— : *Eight Suits of Lessons for the Harpsichord or Spinnet* (London, 1728/*R*1986)
—— : *Voluntarys and Fugues Made on Purpose for the Organ or Harpsicord* (London, *c*1730/*R*1985)
Rossi, M.: *Collected Keyboard Works*, ed. J. R. White, CEKM, xv (1966)
Santa María, Tomás de: *Libro llamado Arte de tañer fantasia* (1545), facs., introduction by D. Stevens (London, 1972)
Scarlatti, A.: *Primo e secondo libro di toccate*, ed. R. Gerlin, CMI, xiii (1943)
—— : [29] *Toccate per cembalo*, ed. J. S. Shedlock (London, 1908–10)
Scarlatti, D.: [555] *Sonates*, ed. K. Gilbert, Le pupitre, xxxi–xli (Paris, 1971–84)
Scheidemann, H.: *Orgelwerke*, ed. G. Fock and W. Breig, 3 vols. (Kassel, 1967–71)
Scheidt, S.: *Tabulatura nova, Hamburg, 1624, Teil I–III*, Samuel Scheidts Werke, vi, vii, ed. C. Mahrenholz (Hamburg, 1953)
Schmid, B. (the elder): Pieces ed. in W. Merian, *Der Tanz* (see Bibliography, 'Repertory', §4)
Seixas, C.: 80 sonatas, ed. M. S. Kastner, PM, x (1965)
Soler, A.: *Sonatas para instrumentos de tecla*, i–vii, ed. S. Rubio (Madrid, 1957–72)
—— : 14 sonatas, ed. K. Gilbert (London, 1987)

Sweelinck, J. P.: *Opera omnia*, i/1–3: *Keyboard works*, ed. G. Leonhardt, UVNM (Amsterdam, 1968)
Tallis, T.: *Complete Keyboard Works*, ed. D. Stevens (London, 1953)
Tomkins, T.: *Keyboard Music*, ed. S. Tuttle, MB, v (1955, 2/1964)
Trabaci, G. M.: 12 pieces ed. in MMI, 1st ser., iii (1964)
Valente, A.: *Intavolatura de Cimbalo* (1576), ed. C. Jacobs (Oxford, 1973)
Zipoli, D.: *Orgel- und Cembalowerke*, i–ii, ed. L. F. Tagliavini (Heidelberg, 1959) [harpsichord works in vol.ii]

4. 20th-century composers

Carter, E.: Double Concerto, for harpsichord, piano and 2 chamber orchestras (New York, 1964)
Delius, F.: *Dance for Harpsichord* (Vienna, 1922)
Falla, M. de: Concerto, for harpsichord with flute, oboe, clarinet, violin and cello (Paris, 1928)
Howells, H.: *Howells' Clavichord* (London, 1961)
—— : *Lambert's Clavichord* (London, 1927)
Leigh, W.: Concertino, for harpsichord and strings (London, 1949)
Martin, F.: Harpsichord Concerto (Vienna, 1969)
—— : *Petite symphonie concertante*, for harpsichord, piano, harp and strings (Vienna, 1951)
Piston, W.: Sonatina, for violin and harpsichord (New York, 1948)
Poulenc, F.: *Concert champêtre*, for harpsichord and orchestra (Paris, 1931)

Bibliography

HARPSICHORD

G. Anselmi: *De musica*, 1434; ed. G. Massera (Florence, 1961)

S. Virdung: *Musica getutscht* (Basle, 1511; repr. 1931); ed. E.K. Niemöller, DM, 1st ser., xxxi (1970)

G. M. Lanfranco: *Scintille di musica* (Brescia, 1533/*R*1970; Eng. trans. in B. Lee: *Giovanni Maria Lanfranco's 'Scintille di Musica' and its Relation to 16th-Century Music Theory* (diss., Cornell U., 1961)

G. Zarlino: *Le istitutioni harmoniche* (Venice, 1558/*R*1965, rev. 3/1573/*R*1966)

L. Fioravanti: *Dello specchio di scientia universale* (Venice, 1564), 273 [on Trasuntino]

V. Galilei: *Dialogo della musica antica et della moderna* (Florence, 1581/*R*1968)

T. Garzoni: *Piazze universale di tutte le professioni del mondo*, Discorso 136 (Venice, 1585) [on Trasuntino]

G. Diruta: *Il transilvano dialogo sopra il vero modo di sonar organi, et istromenti da penna* (Venice, 1593–1609/*R*1978)

A. Banchieri: *L'organo suonarino* (Venice, 2/1611/*R*)

F. Colonna: *La Sambuca lincea overo dell'istromento musico perfetto* (Naples, 1618)

M. Praetorius: *Syntagma musicum*, ii (Wolfenbüttel, 1618, 2/1619/*R*1958 and 1980)

B. Jobernadi: *Tratado de la musica*, 1634 [Biblioteca Nacional, Madrid, MS 8931]; extracts in *AnM*, viii (1953), 193

M. Mersenne: *Harmonie universelle* (Paris, 1636–7/*R*1963; Eng. trans., 1957)

J. Denis: *Traité de l'accord de l'espinette avec la comparaison de son clavier avec la musique vocale* (Paris, 1643, 2/1650/*R*1969)

A. Kircher: *Phonurgia nova* (Kempten, 1673/*R*1966)

M. Todini: *Dichiaratione della Galleria Armonica* (Rome, 1676)

C. Douwes: *Grondig ondersoek van de toonen der musijk* (Franeker, 1699/*R*1970)

J. G. Walther: *Musicalisches Lexicon, oder Musicalische Bibliothec* (Leipzig, 1732); ed. R. Schaal, DM, 1st ser., iii (1953)

J. Adlung: *Musica mechanica organoedi* (Berlin, 1768/*R*1961)

C. Burney: *The Present State of Music in France and Italy* (London, 1771, 2/1773); ed. P. Scholes as *Dr. Burney's Musical Tours*, i (London, 1959)

D. Diderot: 'Clavecin', *Encyclopédie*, iii (Geneva, 1772), 509
C. Burney: *The Present State of Music in Germany, The Netherlands, and the United Provinces* (London, 1773, 2/1775); ed. P. Scholes as *Dr. Burney's Musical Tours*, ii (London, 1959)
A. F. N. Blanchet: *Méthode abrégée pour accorder le clavecin et le piano* (Paris, 1797–1800/*R*1976)
L. de Burbure: *Recherches sur les facteurs de clavecins et les luthiers d'Anvers, depuis le XVIe jusqu'au XIXe siècle* (Brussels, 1869)
G. Correr: *Elenco degli strumenti musicali antichi da arco, fiato, pizzico e tasto* (Venice, 1872)
V. Mahillon: *Catalogue descriptif et analytique du Musée instrumental du Conservatoire royal de musique de Bruxelles* (Ghent, 1880–1922/*R*; i, 2/1893; ii, 2/1909)
C. Krebs: 'Die besaiteten Klavierinstrumente bis zu Anfang des 17. Jahrhunderts', *VMw*, viii (1892), 92, 288; ix (1893), 245
M. Steinert: *Catalogue of the M. Steinert Collection of Keyed and Stringed Instruments* (New Haven, 1893)
A. J. Hipkins: *A Description and History of the Pianoforte and of the Older Keyboard Instruments* (London, 1896, 3/1929/*R*1975)
G. Donaldson: *Catalogue of the Musical Instruments and Objects forming the Donaldson Museum* (London, 1899)
O. Fleischer: 'Das Bach'sche Clavicymbel und seine Neukonstruktion', *ZIMG*, i (1899–1900), 161
K. Nef: 'Clavicymbel und Clavichord', *JbMP 1903*, 15
——: *Katalog der Musikinstrumente im Historischen Museum zu Basel* (Basle, 1906)
——: 'Zur Cembalofrage', *ZIMG*, x (1908–9), 236
J. Schlosser: *Die Sammlung alter Musikinstrumente* (Vienna, 1920)
C. Sachs: *Sammlung alter Musikinstrumente bei der Staatlichen Hochschule für Musik zu Berlin* (Berlin, 1922)
P. Macquoid and R. Edwards: *The Dictionary of English Furniture* (New York, 1924–7, 2/1954)
E. U. Kropp: *Das Zupfklavier* (diss., U. of Berlin, 1925)
P. James: *Early Keyboard Instruments* (London, 1930/*R*1970)
G. Le Cerf and E.-R. Labande, eds.: *Instruments de musique du XVe siècle: les traités d'Henri-Arnaut de Zwolle et de divers anonymes (Ms. B. N. latin 7295)* (Paris, 1932/*R*1972)
K. Matthaei: 'Über Cembalo-Neukonstruktionen', *Zeitschrift für Hausmusik*, ii (1933)
H. Neupert: *Das Cembalo* (Kassel, 1933, 4/1969; Eng. trans., 1960 as *Harpsichord Manual*)
W. Skinner: *The Belle Skinner Collection of Musical Instruments* (Holyoke, 1933)
E. Harich-Schneider: *Die Kunst des Cembalospiels* (Kassel, 1939)
F. Trendelenburg, E. Thienhaus and E. Franz: 'Zur Klangwirkung von

Klavichord, Cembalo und Flügel', *Akustische Zeitschrift*, v (1940), 309
H.-H. Dräger: 'Anschlagsmöglichkeiten beim Cembalo', *AMf*, vi (1941), 223
A. M. Pols: *De Ruckers en de klavierbouw en Vlaanderen* (Antwerp, 1942)
J. A. Stellfeld: *Bronnen tot de geschiedenis der Antwerpsche clavecimbel- en orgelbouwers in de XVI en XVII eeuwen* (Antwerp, 1942)
E. Halfpenny: 'Shudi and the "Venetian Swell"', *ML*, xxvii (1946), 180
J. Wörsching: *Die historischen Saitenklaviere und der moderne Klavichord- und Cembalobau* (Mainz, 1946)
W. Landowska: *Commentaries for the 'Treasury of the Harpsichord Music'* (New York, 1947)
A. Mendel: 'Pitch in the 16th and Early 17th Centuries', *MQ*, xxxiv (1948), 28, 199, 336, 575
N. Dufourcq: *Le clavecin* (Paris, 1949, 2/1967)
A. Mendel: 'Devices for Transposition in the Organ before 1600', *AcM*, xxi (1949), 24; repr. in A. J. Ellis and A. Mendel: *Studies in the History of Musical Pitch* (Amsterdam, 1968), 170
F. Hubbard: 'Two Early English Harpsichords', *GSJ*, iii (1950), 12
A. Berner: *Die Berliner Musikinstrumentensammlung: Einführung mit historischen und technischen Erläuterungen* (Berlin, 1952)
E. Harich-Schneider: *Kleine Schule des Cembalospiels* (Kassel, 1952; Eng. trans., 1954, as *The Harpsichord*)
R. Kirkpatrick: *Domenico Scarlatti* (Princeton, 1953, 5/1966)
J. Boston: 'An Early Virginal-maker in Chester, and his Tools', *GSJ*, vii (1954), 3
F. Ernst: *Der Flügel Johann Sebastian Bachs* (Frankfurt am Main, 1955)
F. J. Hirt: *Meisterwerke des Klavierbaues* (Olten, 1955; Eng. trans., 1968, 2/1981, as *Stringed Keyboard Instruments 1440–1880*)
R. Russell: 'The Harpsichord since 1800', *PRMA*, lxxxii (1955–6), 61
D. H. Boalch: *Makers of the Harpsichord and Clavichord 1440–1840* (London, 1956, 2/1974; rev. 3 in preparation)
F. Hubbard: 'The *Encyclopédie* and the French Harpsichord', *GSJ*, ix (1956), 37
P. J. Hardouin and F. Hubbard: 'Harpsichord Making in Paris', *GSJ*, x (1957), 10; xii (1959), 73; xiii (1960), 52
R. Russell: *Catalogue of the Benton Fletcher Collection of Early Keyboard Instruments* (London, 1957, rev. 2/1969 by H. Schott)
—— : *The Harpsichord and Clavichord* (London, 1959, rev. 2/1973)
A. Curtis: 'Dutch Harpsichord Makers', *TVNM*, xix/1–2 (1960), 44
S. Marcuse: *Musical Instruments at Yale: a Selection of Western Instruments from the 15th to 20th Centuries* (New Haven, 1960)
J. D. Shortridge: 'Italian Harpsichord Building in the 16th and 17th Centuries', *Smithsonian Institution Bulletin*, no.225 (1960, 2/1970)

E. Winternitz: *Keyboard Instruments in the Metropolitan Museum of Art* (New York, 1961)
J. Lade: 'Modern Composers and the Harpsichord', *Consort*, xix (1962), 128
F. Hubbard: *Harpsichord Regulating and Repairing* (Boston, Mass., 1963)
J. Barnes: 'Pitch Variations in Italian Keyboard Instruments', *GSJ*, xviii (1965), 110
F. Hubbard: *Three Centuries of Harpsichord Making* (Cambridge, Mass., 1965, 4/1972)
J. Barnes: 'Two Rival Harpsichord Specifications', *GSJ*, xix (1966), 49
V. Luithlen and K. Wegerer: *Katalog der Sammlung alter Musikinstrumente*, i: *Saitenklaviere* (Vienna, 1966)
J. H. van der Meer: *Beiträge zum Cembalobau im deutschen Sprachgebiet bis 1700* (Nuremberg, 1966)
L. Cervelli and J. H. van der Meer: *Conservato a Roma il più antico clavicembalo tedesco* (Rome, 1967)
E. M. Ripin: 'The Early Clavichord', *MQ*, liii (1967), 518
—— : 'The French Harpsichord before 1650', *GSJ*, xx (1967), 43
W. R. Thomas and J. J. K. Rhodes: 'The String Scales of Italian Keyboard Instruments', *GSJ*, xx (1967), 48
E. Winternitz: *Musical Instruments and their Symbolism in Western Art* (New York, 1967, 2/1979)
J. Barnes: 'Italian String Scales', *GSJ*, xxi (1968), 179
J. H. van der Meer: 'Harpsichord Making and Metallurgy: a Rejoinder', *GSJ*, xxi (1968), 175
C. Mould: 'James Talbot's Manuscript (Christ Church Library Music Manuscript 1187), vii: Harpsichord', *GSJ*, xxi (1968), 40
S. Newman and P. Williams: *The Russell Collection and other Early Keyboard Instruments in St Cecilia's Hall, Edinburgh* (Edinburgh, 1968)
I. Otto: *Das Musikinstrumenten-Museum Berlin* (Berlin, 1968)
E. M. Ripin: 'The Two-manual Harpsichord in Flanders before 1650', *GSJ*, xxi (1968), 33
R. Russell: *Victoria and Albert Museum: Catalogue of Musical Instruments,* i: *Keyboard Instruments* (London, 1968, rev. 2/1985 by H. Schott)
G. Gábry: *Old Musical Instruments* (Budapest, 1969, 2/1976)
V. Gai: *Gli instrumenti musicali della corta medicea* (Florence, 1969)
C. Hoover: *Harpsichords and Clavichords* (Washington, DC, 1969)
R. de Maeyer: *Exposition des instruments de musique des XVIème et XVIIème siècles* (Brussels, 1969)
J. H. van der Meer: 'Die klavierhistorische Sammlung Neupert', *Anzeiger des Germanischen Nationalmuseums* (Nuremberg, 1969)
E. M. Ripin: 'The Couchet Harpsichord in the Crosby Brown Collection', *Metropolitan Museum Journal*, ii (1969), 169
W. J. Zuckermann: *The Modern Harpsichord* (New York, 1969)
E. Ripin: 'Expressive Devices applied to the Eighteenth-century Harpsichord', *Organ Yearbook*, i (1970), 65

J.-L. Val: 'Une détermination de la taille des cordes de clavecin employées en France au XVIIIe siècle', *ReM*, lvi (1970), 208

Restauratieproblemen van Antwerpse klavecimbels (Antwerp, 1970)

K. and M. Kaufmann: 'Le clavecin d'Arnaut de Zwolle', *GAM: Bulletin du Groupe d'acoustique musicale*, liv (1971), Feb, p.i

J. H. van der Meer: *Wegweiser durch die Sammlung historischer Musikinstrumente* (Nuremberg, 1971)

N. Meeùs: 'Le clavecin de Johannes Couchet, Anvers, 1646: un moment important de l'histoire du double clavecin en Flandres', *Brussels Museum of Musical Instruments Bulletin*, i (1971), 15

W. D. Neupert: 'Physikalische Aspekte des Cembaloklanges', *Das Musikinstrument*, xx (1971), 857

E. M. Ripin, ed.: *Keyboard Instruments: Studies in Keyboard Organology 1500–1800* (Edinburgh, 1971, 2/1977) [articles by J. Barnes, E. A. Bowles, F. Hellwig, J. Lambrechts-Douillez, G. Leonhardt, J. H. van der Meer and others]

M. Thomas: 'String Gauges of Old Italian Harpsichords', *GSJ*, xxiv (1971), 69

P. Williams: 'Some Developments in Early Keyboard Studies', *ML*, lii (1971), 272

N. Meeùs: 'Bartolomeo Ramos de Pareja et la tessiture des instruments à clavier entre 1450 et 1550', *Revue des archéologues et historiens d'art de Louvain*, v (1972), 148

O. Rindlisbacher: *Das Klavier in der Schweiz* (Berne and Munich, 1972)

J. Barnes: 'The Stringing of Italian Harpsichords', *Der klangliche Aspekt beim Restaurieren von Saitenklavieren*, ed. V. Schwarz (Graz, 1973), 35

M. Castellani: 'A 1593 Veronese Inventory', *GSJ*, xxvi (1973), 15

L. Cervelli: 'Arpicordo: mito di un nome e realtà di uno strumento', *Quadrivium*, xiv (1973), 187

E. M. Ripin: 'A "Five-foot" Flemish Harpsichord', *GSJ*, xxvi (1973), 135

—— : 'The Surviving Oeuvre of Girolamo Zenti', *Metropolitan Museum Journal*, vii (1973), 71

W. R. Thomas and J. J. K. Rhodes: 'Harpsichord Strings, Organ Pipes, and the Dutch Foot', *Organ Yearbook*, iv (1973), 112

K. Bakeman: 'Stringing Techniques of Harpsichord Builders', *GSJ*, xxvii (1974), 95

H. Bédard and J. Lambrechts-Douillez: 'Rapports de restauration', *Brussels Museum of Musical Instruments Bulletin*, iv (1974), 17

J. H. van der Meer: 'Studien zum Cembalobau in Italien', *Festschrift für Ernst Emsheimer* (Stockholm, 1974), 131

G. G. O'Brien: 'The Numbering System of Ruckers Instruments', *Brussels Museum of Musical Instruments Bulletin*, iv (1974), 75

—— : 'The 1764/83 Taskin Harpsichord', *Organ Yearbook*, v (1974), 91

E. M. Ripin: *The Instrument Catalogues of Leopoldo Franciolini* (New York, 1974)

H. Schott: 'The Harpsichord Revival', *Early Music*, ii (1974), 85
M. Skowroneck: 'Das Cembalo von Christian Zell, Hamburg, und seine Restaurierung', *Organ Yearbook*, v (1974), 79
L. F. Tagliavini: 'Considerazioni sulle vicende storiche del corista', *L'organo*, xii (1974), 119
W. R. Thomas and J. J. K. Rhodes; G. G. O'Brien; D. V. D. Brown: 'A Clavichord, a Harpsichord and a Chamber Organ in the Russell Collection, Edinburgh', *Organ Yearbook*, v (1974), 88
M. Campbell: *Dolmetsch: the Man and his Work* (London, 1975)
P. Dumoulin: 'La découverte de bobines de cordes de clavecins du XVIIIe siècle', *RdM*, lxi (1975), 8
F. Hammond: 'Musical Instruments at the Medici Court in the Mid-seventeenth Century', *AnMc*, no.15 (1975), 202
L. Libin: 'A Dutch Harpsichord in the United States', *GSJ*, xxviii (1975), 43
M. Tiella: 'The Archicembalo of Nicola Vicentino', *English Harpsichord Magazine*, i/5 (1975), 134
La facture de clavecin du XVe au XVIIIe siècle: Louvain-la-Neuve 1976 [articles by J. Bosquet, M. K. Kauffmann, J. Lambrechts-Douillez, H. Legros, N. Meeùs, P. Mercier and J. Tournay]
D. Adlam: 'Restoring the Vaudry', *Early Music*, iv (1976), 255
F. Hellwig: 'Strings and Stringing: Contemporary Documents', *GSJ*, xxix (1976), 91
J. Lambrechts-Douillez, ed.: *Ruckers klavecimbels en copieën* (Antwerp, 1977)
M. Lindley: 'Instructions for the Clavier Diversely Tempered', *Early Music*, v (1977), 18
A. and P. Mactaggart: 'Some Problems Encountered in Cleaning two Harpsichord Soundboards', *Studies in Conservation*, xxii (London, 1977), 73
N. Meeùs: 'Renaissance Transposing Keyboards', *FoMRHI Quarterly*, no.6 (1977), 18; no.7 (1977), 16
W. Debenham: *A Description of the Alterations to the 1521 Hieronymus Harpsichord* (London, 1978)
W. R. Dowd: 'A Classification System for Ruckers and Couchet Double Harpsichords', *JAMIS*, iv (1978), 106
S. Germann: 'Regional Schools of Harpsichord Decoration', *JAMIS*, iv (1978), 54–105
A. and P. Mactaggart: 'The Knole Harpsichord: a Reattribution', *GSJ*, xxi (1978), 2
J. H. van der Meer: 'A Contribution to the History of the Clavicytherium', *Early Music*, vi (1978), 247
A. Mendel: 'Pitch in Western Music since 1500 – a Re-examination', *AcM*, l (1978), 1–93; also pubd separately (Kassel, 1979)
C. Page and L. Jones: 'Four more 15th-century Representations of Stringed Keyboard Instruments', *GSJ*, xxxi (1978), 151

D. Esch: 'Die früheste Erwähnung des Clavicymbalum in italienischer Sprache', *AnMc*, no.19 (1979), 37
F. Hammond: 'Girolamo Frescobaldi and a Decade of Music in Casa Barbarini: 1634–1643', *AnMc*, no.19 (1979), 94–124
H. Henkel: *Beiträge zum historischen Cembalobau*, Beiträge zur musikwissenschaftlichen Forschung in der DDR, xi (Leipzig, 1979)
——: *Kielinstrumente*, Musikinstrumenten-Museum der Karl-Marx-Universität Leipzig Katalog, ii (Leipzig, 1979)
A. and P. Mactaggart: 'Tempera and Decorated Keyboard Instruments', *GSJ*, xxxii (1979), 59
G. G. O'Brien: 'Ioannes and Andreas Ruckers: a Quatercentenary Celebration', *Early Music*, vii (1979), 453
W. Strack: 'Christian Gottlob Hubert and his Instruments', *GSJ*, xxxii (1979), 38
W. R. Thomas and J. J. K. Rhodes: 'Harpsichords and the Art of Wire-drawing', *Organ Yearbook*, x (1979), 126
J.-M. Tuchscherer: 'Le clavecin de Donzelague', *La revue du Louvre*, v/6 (1979), 440
D. Alton Smith: 'The Musical Instrument Inventory of Raymund Fugger', *GSJ*, xxxiii (1980), 36
J. Koster: 'The Importance of the Early English Harpsichord', *GSJ*, xxxiii (1980), 45
L. F. Tagliavini: 'Appunti sugli ambiti delle tastiere in Italia dal rinascimento al primo barocco', *Arte nell'Aretino* (Florence, 1980), 26
M. Thomas: 'Harpsichords which have been found recently in France', *English Harpsichord Magazine*, ii/7 (1980), 158
F. Abondance: *Restauration des instruments de musique* (Fribourg, 1981)
G. Haase and D. Krickeberg: *Tasteninstrumente des Museums* (Berlin, 1981)
T. McGeary: 'Harpsichord Mottoes', *JAMIS*, vii (1981), 5–34
E. Nordenfelt-Åberg: 'The Harpsichord in 18th-century Sweden', *Early Music*, ix (1981), 47
G. G. O'Brien: 'Some Principles of 18th-century Harpsichord Stringing and their Application', *Organ Yearbook*, xii (1981), 160
M. Spencer: 'Harpsichord Physics', *GSJ*, xxxiv (1981), 2
M. Tiella: 'Problemi conessi con il restauro degli strumenti musicali', *Didactica del restauro liutario: Premeno 1981*, 7
D. Woolley: 'The Haward Harpsichord at Knole' [actually Hovingham Hall], *English Harpsichord Magazine*, iii/1 (1981), 2
W. Barry: 'The Keyboard Instruments of King Henry VIII', *Organ Yearbook*, xiii (1982), 31
M.-J. Bosschaerts-Eykens: 'Dokumenten betreffende de familie Hagaerts', *Mededelingen van het Ruckers-genootschap*, ii (1982), 11
J. Bran-Ricci and M. Robin: 'Le clavecin par Simon Hagaerts, Anvers, XVIIe siècle', *Mededelingen van het Ruckers-genootschap*, ii (1982), 25

J. Koster: 'A Remarkable Early Flemish Transposing Harpsichord', *GSJ*, xxxv (1982), 45
J. Lambrechts-Douillez: 'Klavecimbelbouwersfamilie Hagaerts', *Mededelingen van het Ruckers-genootschap*, ii (1982), 1
——: 'Stamboom der klavecimbelbouwersfamilie Ruckers-Couchet', *Mededelingen van het Ruckers-genootschap*, i (1982), 10
J. Lester: 'The Musical Mechanisms of Arnaut de Zwolle', *English Harpsichord Magazine*, iii/3 (1982), 35
D. Wraight: 'Considerations on the Categorisation of Italian Harpsichords', *Congress: Premeno 1982*
M.-J. Bosschaerts-Eykens: 'Dokumenten', *Mededelingen van het Ruckers-genootschap*, iii (1983), 9
R. Gug: 'Histoire d'une corde de clavecin hier et aujourd'hui', *Musique ancienne*, no.15 (1983), 5
J. Lambrechts-Douillez, ed.: *Hans en Joannes Ruckers*, Mededelingen van het Ruckers-genootschap, iii (1983)
S. Leschiutta: *Appunti per una bibliografia sul clavicembalo, clavicordo e fortepiano* (Padua, 1983)
——: *Cembalo, spinetta e virginale* (Ancona, 1983)
A. and P. Mactaggart: 'A Royal Ruckers: Decorative and Documentary History', *Organ Yearbook*, xiv (1983), 78
J. H. van der Meer: *Musikinstrumente* (Nuremberg, 1983)
G. G. O'Brien: 'The Authentic Instruments from the Workshops of Hans and Ioannes Ruckers', *Mededelingen van het Ruckers-genootschap*, iii (1983), 37
G. Stradner: *Spielpraxis und Instrumentarium um 1500: dargestellt an Sebastian Virdung's 'Musica Getutscht' (Basel 1511)* (Vienna, 1983)
D. Wraight: 'Il cembalo italiano al tempo di Frescobaldi: problemi relativi alla misurazione delle corde e alla tastiera', *Girolamo Frescobaldi nel IV centenario della nascità: Ferrara 1983*, 375
R. Gug: 'En remontant la filière de Thoiry à Nuremberg', *Musique ancienne*, no.18 (1984), 4; abridged Eng. trans., *FoMRHI Quarterly*, no.45 (1986), 74
J. Lambrechts-Douillez: 'Andreas Ruckers de oude. Andreas Ruckers de jonge', *Mededelingen van het Ruckers-genootschap*, iv (1984), 17
L. W. Martin: 'The Colonna-Stella "Sambuca lincea": an Enharmonic Keyboard Instrument', *JAMIS*, x (1984), 5
G. G. O'Brien: 'The Authentic Instruments from the Workshops of Andreas Ruckers I and Andreas Ruckers II', *Mededelingen van het Ruckers-genootschap*, iv (1984), 61
H. Schott, ed.: *The Historical Harpsichord*, i: *Essays in Honor of Frank Hubbard* (New York, 1984) [articles by W. R. Dowd, F. Hubbard, G. Leonhardt, C. Page and H. Schott]
R. T. Shann: 'Flemish Transposing Harpsichords: an Explanation', *GSJ*, xxxvii (1984), 62

M. Tiella: 'Renaissance and Baroque Musical Instruments and their "Pronuntia"', *Organ Yearbook*, xv (1984), 5

D. Wraight: 'Italian Two-manual Harpsichords', *FoMHRI Quarterly*, no.36 (1984), 19

——: Review of H. Henkel: *Beiträge zum historischen Cembalobau* (Leipzig, 1979), *FoMRHI Quarterly*, no.34 (1984), 70

P. Barbieri: 'Giordano Riccati on the Diameters of Strings and Pipes', *GSJ*, xxxviii (1985), 20

S. Germann: 'The Mietkes, the Margrave and Bach', *Bach, Handel, Scarlatti: Tercentenary Essays*, ed. P. Williams (Cambridge, 1985), 119–48

E. L. Kottick: 'The Acoustics of the Harpsichord: Resonance Curves and Modes', *GSJ*, xxxviii (1985), 55

P. and A. Mactaggart: 'The Colour of Ruckers Lid Papers', *GSJ*, xxxviii (1985), 106

N. Meeùs: 'An Unpublished Letter by Michael Praetorius', *FoMRHI Quarterly*, no.39 (1985), 32

B. Kenyon de Pascual: 'Diego Fernández – Harpsichord-maker to the Spanish Royal Family from 1722–1755 – and his Nephew Julián Fernández', *GSJ*, xxxviii (1985), 35

H. Schott: 'From Harpsichord to Pianoforte: a Chronology and Commentary', *Early Music*, xiii (1985), 28

——: *Victoria and Albert Museum: Catalogue of Musical Instruments*, i: *Keyboard Instruments* (London, 1985) [rev. edn of Russell, 1968]

J. Shortridge: 'Ruckers "Transposing" Double Harpsichords', *FoMRHI Quarterly*, no.40 (1985), 23

D. Wraight: 'The Conservation of Keyboard Instruments', *European Restoration Conference, Fondazione Levi: Venice 1985*

——: 'Nouvelles études sur les clavecins italiens', *Musique ancienne*, no.20 (1985), 67

M.-J. Bosschaerts-Eykens: 'Dokumenten betreffende de familie Couchet', *Mededelingen van het Ruckers-genootschap*, v (1986), 7

J. Lambrechts-Douillez: 'De familie Couchet', *Mededelingen van het Ruckers-genootschap*, v (1986), 1

J. H. van der Meer: 'A Curious Instrument with a Five-octave Compass', *Early Music*, xiv (1986), 397

G. G. O'Brien: 'The Authentic Instruments from the Workshops of Ioannes Couchet and his Sons', *Mededelingen van het Ruckers-genootschap*, v (1986), 45

L. F. Tagliavini and J. H. van der Meer: *Clavicembali e spinette dal XVI al XIX secolo* (Bologna, 1986)

D. Wraight: 'Neue Untersuchungen an italienischen Cembali', *Concerto*, iii/2 (1986), 28

——: 'Vincentius and the Earliest Harpsichords', *Early Music*, xiv (1986), 534

EARLY KEYBOARD

M. Goodway and J. S. Odell: *The Historical Harpsichord*, ii: *The Metallurgy of 17th- and 18th-century Music Wire* (Stuyvesant, NY, 1987)
L. Libin: 'Folding Harpsichords', *Early Music*, xv (1987), 378
P. Mactaggart: 'Examination and Restoration of Paint on Musical Instruments', *Restoration of Early Musical Instruments*, ed. C. Huntley and K. Starling (London, 1987), 6 [UK Institute for Conservation, Occasional Paper, no.6]
C. Nobbs: 'A Seventeenth Century Harpsichord', *Harpsichord and Fortepiano Magazine*, iv/3 (1987), 46
B. Kenyon de Pascual: 'Francisco Pérez Mirabal's Harpsichords and the Early Spanish Piano', *Early Music*, xv (1987), 503
H. Heyde: 'Zum Florentiner Cembalobau um 1700-Bemerkung zu MS-68 und MS–70 des Händel-Hauses Halle', *Festschrift für John Henry Van der Meer zu seinem fünfundsechzigsten Geburtstag*, Wissenschaftliche Beibände zum Anzeiger des Germanischen Nationalmuseums, vi (Tutzing, 1987)
D. Krickeberg and H. Rase: 'Beiträge zur Kenntnis des mittel- und norddeutschen Cembalobaus um 1700', *Festschrift für John Henry Van der Meer* (Tutzing, 1987)
J. Lambrechts-Douillez: 'The History of Harpsichord Making in Antwerpen in the 18th Century', *Festschrift für John Henry Van der Meer* (Tutzing, 1987)
D. Wraight: 'The Celestini Harpsichord of 1605: another Misleading Instrument', *Organ Yearbook* (in preparation)

VIRGINAL AND SPINET

G. Zarlino: *Le istitutioni harmoniche* (Venice, 1558/*R*1965, rev. 3/1573/*R*1966)
A. Banchieri: *L'organo suonarino* (Venice, 2/1611/*R*)
M. Praetorius: *Syntagma musicum*, ii (Wolfenbüttel, 1618, 2/1619/*R*1958 and 1980)
C. Douwes: *Grondig ondersoek van de toonen der musijk* (Franeker, 1699/*R*1970)
D. Boalch: *Makers of the Harpsichord and Clavichord 1440–1840* (London, 1956, 2/1974; rev. 3 in preparation)
R. Russell: *The Harpsichord and Clavichord* (London, 1959, rev. 2/1973)
S. Marcuse: *Musical Instruments: a Comprehensive Dictionary* (New York, 1964), 581
F. Hubbard: *Three Centuries of Harpsichord Making* (Cambridge, Mass., 1965, 4/1972)
V. Gai: *Gli instrumenti musicali della corta medicea* (Florence, 1969)
G. Leonhardt: 'In Praise of Flemish Virginals of the Seventeenth Century', *Keyboard Instruments: Studies in Keyboard Organology 1500–1800*, ed. E. M. Ripin (Edinburgh, 1971, 2/1977)

J. H. van der Meer: 'Beiträge zum Cembalobau der Familie Ruckers', *Jb des Staatlichen Instituts für Musikforschung Preussischer Kulturbesitz* (1971), 100–153
E. M. Ripin: 'On Joes Karest's Virginal and the Origins of the Flemish Tradition', *Keyboard Instruments: Studies in Keyboard Organology 1500–1800* (Edinburgh, 1971, 2/1977)
—— : 'The Surviving Oeuvre of Girolamo Zenti', *Metropolitan Museum Journal*, vii (1973), 71
W. R. Thomas and J. J. K. Rhodes: 'Harpsichord Strings, Organ Pipes, and the Dutch Foot', *Organ Yearbook*, iv (1973), 112
J. H. van der Meer: 'Studien zum Cembalobau in Italien', *Festschrift für Ernst Emsheimer* (Stockholm, 1974), 131, 275
N. Meeùs: 'La facture de virginales à Anvers du 16e siècle', *Brussels Museum of Musical Instruments Bulletin*, iv (1974), 55
G. G. O'Brien: 'The Numbering System of Ruckers Instruments', *Brussels Museum of Musical Instruments Bulletin*, iv (1974), 75
N. Meeùs: 'Epinettes et "muselars": une analyse théorique', *La facture de clavecin du XVe au XVIIIe siècle': Louvain-la-Neuve 1976*, 67
J. Koster: 'The Mother and Child Virginal and its Place in the Keyboard Instrument Culture of the Sixteenth and Seventeenth Centuries', *Ruckers klavecimbels en copieën: Antwerp 1977*, 78
G. G. O'Brien: 'The Stringing and Pitches of Ruckers Instruments', *Ruckers klavecimbels en copieën: Antwerp 1977*, 48
G. Nitz: *Die Klanglichkeit in der englischen Virginalmusik des 16. Jahrhunderts* (Munich, 1979)
S. Howell: 'Paulus Paulirinus of Prague on Musical Instruments', *JAMIS*, v–vi (1979–80), 9
N. Meeùs: 'The Nomenclature of Plucked Keyboard Instruments', *FoMRHI Quarterly*, no.25 (1981), 18
D. Wraight: 'Il cembalo italiano al tempo di Frescobaldi: problemi relativi alla misurazione delle corde e alla tastiera', *Girolamo Frescobaldi nel IV centenario della nascità: Ferrara 1983*, 375
J. Barnes: *Making a Spinet by Traditional Methods* (Welwyn, Herts., 1985) [with reference to instruments by S. Keene]
L. F. Tagliavini and J. H. van der Meer: *Clavicembali e spinette dal XVI al XIX secolo* (Bologna, 1986)

For further bibliography see § 'Harpsichord' above.

CLAVICHORD

S. Virdung: *Musica getutscht* (Basle, 1511; repr. 1931); ed. E. K. Niemöller, DM, 1st ser., xxxi (1970)
J. Bermudo: *Declaración de instrumentos musicales* (Osuna, 1555/*R*1957)
T. de Santa María: *Arte de tañer fantasia* (Valladolid, 1565/*R*1972)
P. Cerone: *El Melopeo y maestro* (Naples, 1613/*R*1954)

M. Praetorius: *Syntagma musicum*, ii (Wolfenbüttel, 1618, 2/1619/*R*1958 and 1980)
F. Correa de Arauxo: *Facultad organica* (Alcalá, 1626/*R*1948), f.25
M. Mersenne: *Harmonie universelle* (Paris, 1636–7/*R*1963; Eng. trans., 1957)
C. Douwes: *Grondig ondersoek van de toonen der musijk* (Franeker, 1699/*R*1970), 98ff, 119ff
J. Mattheson: *Das neu-eröffnete Orchestre* (Hamburg, 1713), 262ff
P. Nassarre: *Escuela música* (Saragossa, 1723–4), 471ff
Q. van Blankenburg: *Elementa musica* (The Hague, 1739/*R*1973), 146f
C. P. E. Bach: *Versuch über die wahre Art das Clavier zu spielen* (Berlin, 1753–62; Eng. trans., 1949)
J. Adlung: *Anleitung zu der musikalischen Gelahrtheit* (Erfurt, 1758/*R*1953), 568ff; (rev. 2/1783)
J. S. Petri: *Anleitung zur praktischen Musik* (Leipzig, 1767, 2/1782/*R*1969)
J. Adlung: *Musica mechanica organoedi*, ii (Berlin, 1768/*R*1961), 123ff, 143f
P. N. Sprengel: *P. N. Sprengels Handwerke und Künste in Tabellen*, xi (Berlin, 1773), 241ff
J. N. Forkel: *Musikalischer Almanach 1782* (Leipzig, 1781/*R*1974), 19 [on the cembal d'amour]
C. F. D. Schubart: *Musicalische Rhapsodien* (Stuttgart, 1786)
D. G. Türk: *Klavierschule* (Leipzig, 1789/*R*1962)
J. Verschuere-Reynvaan: *Muzijkaal kunst-woordenboek*, i (Amsterdam, 2/1795), 143ff
C. F. D. Schubart: *Ideen zu einer Ästhetik der Tonkunst* (Vienna, 1806/*R*1969), 288f
F. X. Wöber, ed.: *Der Minne Regel von Eberhart Cersne* (Vienna, 1861)
A. J. Hipkins: *A Description and History of the Pianoforte and of the Older Keyboard Instruments* (London, 1896, 3/1929/*R*1975), 57ff
A. Goehlinger: *Geschichte des Klavichords* (Basle, 1910)
O. Kinkeldey: *Orgel und Klavier in der Musik des 16. Jahrhunderts* (Leipzig, 1910/*R*1968)
E. van der Straeten: 'The Cembal d'amour', *MT*, lxv (1924), 40
C. Auerbach: *Die deutsche Clavichordkunst des 18. Jahrhunderts* (Kassel, 1930, 3/1959)
G. Le Cerf and E.-R. Labande, eds.: *Instruments de musique du XVe siècle: les traités d'Henri-Arnaut de Zwolle et de divers anonymes (Ms. B. N. latin 7295)* (Paris, 1932/*R*1972)
C. G. Parrish: *The Early Piano and its Influences on Keyboard Technique and Composition in the Eighteenth Century* (diss., Harvard U., 1939)
H. Neupert: *Das Klavichord* (Kassel, 1948, 2/1955; Eng. trans., 1965)

W. Nef: 'The Polychord', *GSJ*, iv (1951), 20

E. Flade: *Gottfried Silbermann* (Leipzig, 1952), 242ff [on the cembal d'amour]

M. S. Kastner: 'Portugiesische und spanische Clavichorde des 18. Jahrhunderts', *AcM*, xxiv (1952), 52

D. H. Boalch: *Makers of the Harpsichord and Clavichord 1440–1840* (London, 1956, 2/1974; rev. 3 in preparation)

F. Lesure: 'Le traité des instruments de musique de Pierre Trichet', *AnnM*, iv (1956), 235ff; pubd separately (Paris, 1957)

R. Russell: *The Harpsichord and Clavichord* (London, 1959, rev. 2/1973)

T. Dart: 'The Clavichord', *Musical Instruments through the Ages*, ed. A. Baines (Harmondsworth, 1961, rev. 2/1966), 66

E. R. Jacobi, ed.: *D. G. Türk: Klavierschule*, DM, xxiii (1962)

E. M. Ripin: 'The Early Clavichord', *MQ*, liii (1967), 518

L. Wagner: 'The Clavichord Today', *Periodical of the Illinois State Music Teachers Association*, vi/1 (1968), 20; vii/1 (1969), 1

E. M. Ripin: 'A Scottish Encyclopedist and the Piano Forte', *MQ*, lv (1969), 487 [on the cembal d'amour]

—— : 'A Reassessment of the Fretted Clavichord', *GSJ*, xxiii (1970), 40

E. A. Bowles: 'A Checklist of 15th-century Representations of Stringed Keyboard Instruments', *Keyboard Instruments: Studies in Keyboard Organology 1500–1800*, ed. E. M. Ripin (Edinburgh, 1971, 2/1977), 11ff, pll.1–31

K. Cooper: *The Clavichord in the 18th Century* (diss., Columbia U., 1971)

G. Doderer: *Clavicórdios portuguesas do século dezoito* (Lisbon, 1971)

W. R. Thomas and J. J. K. Rhodes: 'A Clavichord, a Harpsichord and a Chamber Organ in the Russell Collection, Edinburgh', *Organ Yearbook*, v (1974), 88

R. W. Burhans: 'Audio Engineering Improvements for Clavichords', *Journal of the Audio Engineering Society*, xxiii (1975), 635

J. H. van der Meer: 'The Dating of German Clavichords', *Organ Yearbook*, vi (1975), 100

E. M. Ripin: 'Towards an Identification of the Chekker', *GSJ*, xxviii (1975), 11

B. B. Hoag: 'A Spanish Clavichord Tuning of the 17th Century', *JAMIS*, ii (1976), 86

S. L. Vodraska: 'The Flemish Octave Clavichord: Structure and Fretting', *Organ Yearbook*, x (1979), 117

H. Henkel: *Clavichorde*, Musikinstrumenten-Museum der Karl-Marx-Universität Leipzig Katalog, iv (Leipzig, 1981)

S. Thwaites: 'Some Acoustics of a Clavichord', *Catgut Acoustical Society Newsletter*, no.35 (1981), 29

T. McGeary: 'David Tannenberg and the Clavichord in Eighteenth-century America', *Organ Yearbook*, xiii (1982), 94

RELATED INSTRUMENTS

B. Ramos de Pareia: *Musica pratica* (Bologna, 1482); ed. J. Wolf (Leipzig, 1901/*R*1968)
G. Reisch: *Margarita philosophica* (Freiburg, 1503)
S. Virdung: *Musica getutscht* (Basle, 1511; repr. 1931); ed. E. K. Niemöller, DM, 1st ser., xxxi (1970)
G. M. Lanfranco: *Scintille di musica* (Brescia, 1533/*R*1970; Eng. trans. in B. Lee: *Giovanni Maria Lanfranco's 'Scintille di Musica' and its Relation to 16th-Century Music Theory* (diss., Cornell U., 1961)
V. Galilei: *Dialogo della musica antica et della moderna* (Florence, 1581/*R*1968)
A. Banchieri: *L'organo suonarino* (Venice, 1605, 2/1611/*R*)
R. Cotgrave: *Dictionarie of the French and English Tongues* (London, 1611/*R*1950)
M. Praetorius: *Syntagma musicum*, ii (Wolfenbüttel, 1618, 2/1619/*R*1958 and 1980)
M. Mersenne: *Harmonicorum libri XII* (Paris, 1648, 2/1652)
A. Kircher: *Musurgia universalis* (Rome, 1650/*R*1970)
M. Todini: *Dichiaratione della Galleria Armonica* (Rome, 1676)
F. Bonanni: *Gabinetto armonico pieno d'istromenti sonori indicati e spiegati* (Rome, 1722), 90
J. Mattheson: *Der vollkommene Capellmeister* (Hamburg, 1739/*R*1954)
F. Bédos de Celles: *L'art du facteur d'orgues* (Paris, 1766–78/*R*1963–6; Eng. trans., 1977)
J. Adlung: *Musica mechanica organoedi*, ii (Berlin, 1768/*R*1961)
C. Burney: *The Present State of Music in France and Italy* (London, 1771, 2/1773); ed. P. Scholes as *Dr. Burney's Musical Tours*, i (London, 1959)
C. Engel: *A Descriptive Catalogue of the Musical Instruments in the South Kensington Museum, preceded by an Essay on the History of Musical Instruments* (London, 1874)
E. vander Straeten: *La musique aux Pays-Bas avant le XIXe siècle*, vii (Brussels, 1885/*R*1969)
A. J. Hipkins: *A Description and History of the Pianoforte and of the Older Keyboard Instruments* (London, 1896, 3/1929/*R*1975)
F. Pedrell: *Emporio científico e histórico de organografía musical española antigua* (Barcelona, 1901), 64ff
G. Kinsky: 'Hans Haiden: der Erfinder der Nürnbergischen Geigenwerks', *ZMw*, vi (1923–4), 193
A. Pirro: *Les clavecinistes: étude critique* (Paris, 1924), 5ff
J. Reiss: 'Pauli Paulirini de Praga "Tractatus de Musica" (etwa 1460)', *ZMw*, vii (1924–5), 259
W. H. Flood: 'The Eschequier Virginal: an English Invention', *ML*, vi (1925), 151

BIBLIOGRAPHY

H. G. Farmer: 'The Canon and Eschaquiel of the Arabs', *Journal of the Royal Asiatic Society* (1926), 252

H. Herrmann: *Die Regensburger Klavierbauer Späth und Schmahl und ihr Tangentenflügel* (Erlangen, 1928)

G. Le Cerf and E.-R. Labande, eds.: *Instruments de musique du XVe siècle: les traités d'Henri-Arnaut de Zwolle et de divers anonymes (Ms. B. N. latin 7295)* (Paris, 1932/*R*1972)

R. E. M. Harding: *The Piano-forte: its History Traced to the Great Exhibition of 1851* (Cambridge, 1933, rev.2/1978)

C. Sachs: *The History of Musical Instruments* (New York, 1940)

E. Halfpenny: 'The Lyrichord', *GSJ*, iii (1950), 46

F. W. Galpin: 'Chekker', *Grove 4*, suppl.

D. H. Boalch: *Makers of the Harpsichord and Clavichord 1440–1840* (London, 1956, 2/1974; rev. 3 in preparation)

L. Cervelli: 'Italienische Musikinstrumente in der Praxis des Generalbaßspiels: das Arpichord', *IMSCR*, vii *Cologne 1958*, 76

R. Russell: *The Harpsichord and Clavichord* (London, 1959, rev. 2/1973)

J. H. van der Meer: 'Zur Geschichte des Klaviziteriums', *GfMKB, Kassel 1962*, 305

S. Marcuse: *Musical Instruments: a Comprehensive Dictionary* (New York, 1964)

F. Hubbard: *Three Centuries of Harpsichord Making* (Cambridge, Mass., 1965, 4/1972)

J. H. van der Meer: *Beiträge zum Cembalobau im deutschen Sprachgebiet bis 1700*, Anzeiger des Germanischen Nationalmuseums (Nuremberg, 1966), 103–34

H. Ferguson: 'Bach's "Lautenwerck"', *ML*, xlviii (1967), 259

A. Neven: 'L'arpicordo', *AcM*, xlii (1970), 230

E. A. Bowles: 'A Checklist of Fifteenth-century Representations of Stringed Keyboard Instruments', *Keyboard Instruments: Studies in Keyboard Organology 1500–1800*, ed. E. M. Ripin (Edinburgh, 1971, 2/1977), 11ff, pll.1–31

F. J. de Hen: 'The Truchado Instrument: a Geigenwerk?', *Keyboard Instruments: Studies in Keyboard Organology 1500–1800*, ed. E. M. Ripin (Edinburgh, 1971, 2/1977)

P. Williams: 'The Earl of Wemyss' Claviorgan and its Context in Eighteenth-century England', *Keyboard Instruments: Studies in Keyboard Organology 1500–1800*, ed. E. M. Ripin (Edinburgh, 1971, 2/1977)

L. Cervelli: 'Arpicordo: mito di un nome e realtà di uno strumento', *Quadrivium*, xiv (1973), 187

J. H. van der Meer: 'Studien zum Cembalobau in Italien', *Festschrift für Ernst Emsheimer* (Stockholm, 1974), 131

S. Marcuse: *A Survey of Musical Instruments* (London, 1975), 308

E. M. Ripin: 'Towards an Identification of the Chekker', *GSJ*, xxviii (1975), 11

D. J. Hamoen: 'The Arpicordo Problem: Armand Neven's Solution Reconsidered', *AcM*, xlviii (1976), 181

J. Tournay: 'A propos d'Albertus Delin (1712–1771): petite contribution à l'histoire du clavecin', *La facture de clavecin du XVe au XVIIIe siècle: Louvain-la-Neuve 1976*, 139

J. H. van der Meer: 'Das Arpicordo-Problem nochmals erörtert', *AcM*, xlix (1977), 275

W. Debenham: 'The Compass of the Royal College of Music Clavicytherium', *FoMRHI Quarterly*, no.11 (1978), 19

J. H. van der Meer: 'A Contribution to the History of the Clavicytherium', *Early Music*, vi (1978), 247

C. Page and L. Jones: 'Four More 15th-century Representations of Stringed Keyboard Instruments', *GSJ*, xxxi (1978), 151

E. M. Ripin: 'En Route to the Piano: a Converted Virginal', *Metropolitan Museum Journal*, xiii (1978), 79

E. P. Wells: 'The London Clavicytherium', *Early Music*, vi (1978)

C. Page: 'The Myth of the Chekker', *Early Music*, vii (1979), 482

T. Knighton: 'Another Chekker Reference', *Early Music*, viii (1980), 375

U. Henning: 'The Most Beautiful among the Claviers', *Early Music*, x (1982), 477

W. Barry: 'Preliminary Guidelines for a Classification of Claviorgana', *Organ Yearbook*, xv (1984), 98

E. M. Ripin: 'Chekker', *Grove I*

W. Barry: 'Henri Arnault de Zwolle's *Clavichordium* and the Origin of the Chekker', *JAMIS*, xi (1985), 5

N. Meeùs: 'The Chekker', *Organ Yearbook*, xvi (1985), 5

L. F. Tagliavini and J. H. van der Meer: *Clavicembali e spinette dal XVI al XIX secolo* (Bologna, 1986)

REPERTORY

1. General

A. Méreaux: *Les clavecinistes de 1637 à 1790* (Paris, 1867)

R. Eitner: 'Tänze des 15. bis 17. Jahrhunderts', *MMg*, vii (1875), suppl.

J. W. von Wasielewski: *Geschichte der Instrumentalmusik im 16. Jahrhundert* (Berlin, 1878)

A. G. Ritter: *Zur Geschichte des Orgelspiels, vornehmlich des deutschen, im 14. bis zum Anfange des 18. Jahrhunderts* (Leipzig, 1884/*R*1969)

M. Seiffert: 'J. P. Sweelinck und seine direkten deutschen Schüler', *VMw*, vii (1891), 145–260

—— : *Geschichte der Klaviermusik*, i (Leipzig, 1899/*R*1966)

T. Norlind: 'Zur Geschichte der Suite', *SIMG*, vii (1905–6), 172–203

O. Kinkeldey: *Orgel und Klavier in der Musik des 16. Jahrhunderts* (Leipzig, 1910/*R*1968)

BIBLIOGRAPHY

J. Wolf: *Handbuch der Notationskunde* (Leipzig, 1913–19/*R*1963)
A. Schering: *Studien zur Geschichte der Frührenaissance* (Leipzig, 1914)
J. Müller-Blattau: *Grundzüge einer Geschichte der Fuge* (Königsberg, 1923, 2/1931)
I. Faisst: 'Beiträge zur Geschichte der Claviersonate von ihrem ersten Auftreten an bis auf C. Ph. Bach', *NBJb*, i (1924), 7–85
A. Pirro: *Les clavecinistes* (Paris, 1924)
L. Schrade: 'Ein Beitrag zur Geschichte der Tokkata', *ZMw*, viii (1925–6), 610
O. Deffner: *Über die Entwicklung der Fantasie für Tasten-Instrumente bis J. P. Sweelinck* (diss., U. of Kiel, 1927; Kiel, 1928)
L. Schrade: *Die ältesten Denkmäler der Orgelmusik als Beitrag zu einer Geschichte der Toccata* (Münster, 1928)
K. G. Fellerer: *Orgel und Orgelmusik: ihre Geschichte* (Augsburg, 1929)
R. Gress: *Die Entwicklung der Klavier-Variation von Andrea Gabrieli bis zu Johann Sebastian Bach* (Augsburg, 1929)
Y. Rokseth: *La musique d'orgue au XVe siècle et au début du XVIe* (Paris, 1930)
E. Valentin: *Die Entwicklung der Tokkata im 17. und 18. Jarhhundert (bis J. S. Bach)* (Münster, 1930)
J. Wolf: *Musikalische Schrifttafeln* (Bückeburg, 1930)
L. Schrade: *Die handschriftliche Überlieferung der ältesten Instrumental-Musik* (Lahr, 1931/*R*1968)
H. Klotz: *Über die Orgelkunst der Gotik, der Renaissance und des Barock* (Kassel, 1934)
A. Schering: 'Zur Alternatim-Orgelmesse', *ZMw*, xvii (1935), 19
G. Frotscher: *Geschichte des Orgel-Spiels und der Orgel-Komposition* (Berlin, 1935–6, enlarged 3/1966)
W. Apel: 'Neapolitan Links between Cabezón and Frescobaldi', *MQ*, xxiv (1938), 419
E. Hutchinson: *The Literature of the Piano* (New York, 1938, rev. 3/1964)
E. Epstein: *Der französische Einfluss auf die deutsche Klavier-Suite im 17. Jahrhundert* (Würzburg, 1940)
J. L. Hibberd: *The Early Keyboard Prelude* (diss., Harvard U., 1940)
G. Reese: *Music in the Middle Ages* (New York, 1940)
G. Schünemann: *Geschichte der Klaviermusik* (Berlin, 1940)
W. Georgii: *Klaviermusik* (Zurich and Freiburg, 1941, 4/1965)
O. Gombosi: 'About Dance and Dance Music in the Late Middle Ages', *MQ*, xxvii (1941), 289
W. Apel: *The Notation of Polyphonic Music 900–1600* (Cambridge, Mass., 1942, 5/1953/*R*1961; Ger. trans., 1962)
——: *Masters of the Keyboard* (Cambridge, Mass., 1947/*R*1952)
M. F. Bukofzer: *Music in the Baroque Era* (New York, 1947)
W. Apel: 'Early History of the Organ', *Speculum*, xxiii (1948), 191
R. U. Nelson: *The Technique of Variation* (Berkeley, 1948)

W. Apel: 'The Early Development of the Organ Ricercar', *MD*, iii (1949), 139
G. S. Bedbrook: *Keyboard Music from the Middle Ages to the Beginnings of the Baroque* (London, 1949/*R*1973, with introduction by F. E. Kirby)
M. Kenyon: *Harpsichord Music* (London, 1949)
R. Lunelli: 'Contributi trentini alle relazioni musicali fra l'Italia e la Germania nel Rinascimento', *AcM*, xxi (1949), 41
H. H. Eggebrecht: 'Terminus "Ricercar"', *AMw*, ix (1952), 137
M. S. Kastner: 'Parallels and Discrepancies between English and Spanish Keyboard Music of the Sixteenth and Seventeenth Century', *AnM*, vii (1952), 77–115
W. S. Newman: 'A Checklist of the Earliest Keyboard "Sonatas"', *Notes*, xi (1953–4), 201
J. Friskin and I. Freundlich: *Music for the Piano . . . from 1580 to 1952* (New York, 1954/*R*1973)
H. Hering: 'Das Tokkatische', *Mf*, vii (1954), 277
L. Hoffmann-Erbrecht: *Deutsche und italienische Klavier-Musik zur Bach-Zeit* (Leipzig, 1954)
M. S. Kastner: 'Rapports entre Schlick et Cabezón', *La musique instrumentale de la Renaissance: CNRS Paris 1954*, 217
R. Murphy: *Fantasia and Ricercare in the Sixteenth Century* (diss., Yale U., 1954)
G. Reese: *Music in the Renaissance* (New York, 1954, rev. 2/1959)
A. E. F. Dickinson: 'A Forgotten Collection' [Deutsche Staatsbibliothek, Berlin, Ly.A1 and A2], *MR*, xvii (1956), 97
K. von Fischer: 'Chaconne und Passacaglia', *RBM*, xii (1958), 19
C. Parrish: *The Notation of Medieval Music* (London, 1958)
I. Horsley: 'The 16th-century Variation: a New Historical Survey', *JAMS*, xii (1959), 118
W. S. Newman: *The Sonata in the Baroque Era* (Chapel Hill, 1959, rev. 4/1983)
W. Apel: 'Drei plus drei plus zwei = vier plus vier', *AcM*, xxxii (1960), 29
W. Apel and K. von Fischer: 'Klaviermusik', *MGG*
F. W. Riedel: *Quellenkundliche Beiträge zur Geschichte der Musik für Tasteninstrumente in der zweiten Hälfte des 17. Jahrhunderts* (Kassel, 1960)
Y. Rokseth: 'The Instrumental Music of the Middle Ages and Early 16th Century', *NOHM*, iii (1960), 406–65
H. Alker: *Literatur für alte Tasteninstrumente: Wiener Abhandlungen zur Musikwissenschaft und Instrumentalkunde* (Vienna, 1962)
T. Dart: 'Elisabeth Eysbock's Keyboard Book', *STMf*, xliv (1962), 5; also in *Hans Albrecht in memoriam* (Kassel, 1962), 84
W. Young: 'Keyboard Music to 1600', *MD*, xvi (1962), 115–50; xvii (1963), 163–93

R. S. Douglass: *The Keyboard Ricercar in the Baroque Era* (diss., U. of North Texas, 1963)

E. Southern: 'Some Keyboard Basse Dances of the Fifteenth Century', *AcM*, xxxv (1963), 114

J. Gillespie: *Five Centuries of Keyboard Music* (Berkeley and Los Angeles, 1965/*R*1972)

A. E. F. Dickinson: 'The Lübbenau Keyboard Books' [Deutsche Staatsbibliothek, Berlin, Ly.A1 and A2], *MR*, xxvii (1966), 270

F. E. Kirby: *A Short History of Keyboard Music* (New York, 1966)

J. Caldwell: 'The Organ in the Medieval Latin Liturgy, 800–1500', *PRMA*, xciii (1966–7), 11

W. Apel: *Geschichte der Orgel- und Klaviermusik bis 1700* (Kassel, 1967; Eng. trans., rev. 1972)

H. M. Brown: *Instrumental Music Printed before 1600: a Bibliography* (Cambridge, Mass., 2/1967)

K. Wolters: *Handbuch der Klavierliteratur* (Zurich, 1967)

W. Apel: 'Probleme der Alternierung in der liturgischen Orgelmusik bis 1600', *Congresso internazionale sul tema Claudio Monteverdi e il suo tempo: Venezia, Mantova e Cremona 1968*, 171

—— : 'Solo Instrumental Music', *NOHM*, iv (1968), 602–708

M. C. Bradshaw: *The Origin of the Toccata*, MSD, xxviii (1972)

M. Hinson: *Guide to the Pianist's Repertoire*, ed. I. Freundlich (Bloomington, Ind., 1973) [comprehensive bibliography]

F. Bedford and R. Conant: *Twentieth-century Harpsichord Music* (Hackensack, NJ, 1974)

H. Ferguson: *Keyboard Interpretation from the Fourteenth to the Nineteenth Century: an Introduction* (London, 1975, rev. 2/1987)

N. Bergenfeld: *The Keyboard Fantasy of the Elizabethan Renaissance* (diss., U. of New York, 1978)

C. Rosen: *Sonata Forms* (New York, 1980)

P. le Huray, ed.: *Fingering of Virginal Music* (London, 1981) [anthology of fingerings from virginalist sources]

P. Caldwell: 'Keyboard Music: 1630–1700', *NOHM*, vi (1986), 505–89

W. Emery: 'Organ Music: 1700–1750, *NOHM*, vi (1986), 650–93

P. Radcliffe: 'Harpsichord Music: 1700–1750', *NOHM*, vi (1986), 590–649

2. Italy

L. Torchi: *La musica strumentale in Italia nei secoli XVI, XVII e XVIII* (Turin, 1901)

A. Sandberger: 'Zur älteren italienischen Klaviermusik', *JbMP 1918*, 17

G. Pannain: *Le origini e lo sviluppo dell'arte pianistica in Italia dal 1500 al 1730 circa* (Naples, 1919)

L. Schrade: 'Tänze aus einer anonymen italienischen Tabulatur', *ZMw*, x (1927–8), 449

K. G. Fellerer: 'Zur italienischen Orgelmusik des 17./18. Jahrhunderts', *JbMP 1937*, 70
K. Jeppesen, ed.: *Die italienische Orgelmusik am Anfang des Cinquecento* (Copenhagen, 1943; rev., enlarged 2/1960)
F. Morel: *Girolamo Frescobaldi* (Winterthur, 1945)
D. Plamenac: 'Keyboard Music of the 14th Century in Codex Faenza 117', *JAMS*, iv (1951), 179
A. Machabey: *Girolamo Frescobaldi Ferrarensis* (Paris, 1952)
D. Plamenac: 'New Light on Codex Faenza 117', *IMSCR*, v *Utrecht 1952*, 310
C. Sartori: *Bibliografia della musica strumentale italiana stampata in Italia fino al 1700* (Florence, 1952–68)
N. Pirrotta: 'Note su un codice di antiche musiche per tastiera' [Codex Faenza 117], *RMI*, lvi (1954), 333
S. Podolsky: *The Variation Canzona for Keyboard Instruments in Italy, Austria and Southern Germany in the Seventeenth Century* (diss., Boston U., 1954)
K. Jeppesen: 'Eine frühe Orgelmesse aus Castell'Arquato', *AMw*, xii (1955), 187
M. S. Kastner: 'Una intavolatura d'organo italiana del 1598', *CHM*, ii (1957), 237
L. F. Tagliavini: 'Un musicista cremonese dimenticato', *CHM*, ii (1957), 413
R. Lunelli: *L'arte organaria del Rinascimento in Roma* (Florence, 1958)
W. E. McKee: *The Music of Florientio Maschera (1540–1584)* (diss., North Texas State College, 1958)
B. Becherini: *Catalogo dei manoscritti musicali della Biblioteca nazionale di Firenze* (Kassel, 1959)
J. F. Monroe: *Italian Keyboard Music in the Interim between Frescobaldi and Pasquini* (diss., U. of North Carolina, 1959)
W. Apel: 'Tänze und Arien für Klavier aus dem Jahre 1588', *AMw*, xvii (1960), 51
U. Prota-Giurleo: 'Giovanni Maria Trabaci e gli organisti della Real cappella di palazzo di Napoli', *L'organo*, i (1960), 185
O. Mischiati: 'Tornano alla luce i ricercari della "Musica nova" del 1540', *L'organo*, ii (1961), 73
H. C. Slim: *The Keyboard Ricercar and Fantasia in Italy, ca. 1500–1550* (diss., Harvard U., 1961)
W. Apel: 'Die handschriftliche Überlieferung der Klavierwerke Frescobaldis', *Festschrift Karl Gustav Fellerer* (Regensburg, 1962), 40
—— : 'Die süditalienische Clavierschule des 17. Jahrhunderts', *AcM*, xxxiv (1962), 128
K. Jeppesen: 'Ein altvenetianisches Tanzbuch' [Biblioteca Nazionale Marciana, Venice, Ital. IV. 1227], *Festschrift Karl Gustav Fellerer* (Regensburg, 1962), 245

U. Prota-Giurleo: 'Due campioni della scuola napoletana del sec. xvii', *L'organo*, iii (1962), 115
H. C. Slim: 'Keyboard Music at Castell'Arquato by an Early Madrigalist', *JAMS*, xv (1962), 35
S. Kunze: *Die Instrumentalmusik Giovanni Gabrielis* (Tutzing, 1963)
D. Plamenac: 'A Note on the Rearrangement of Fa', *JAMS*, xvii (1964), 78
—— : 'Faventina', *Liber amicorum Charles van den Borren* (Antwerp, 1964), 145
H. B. Lincoln: 'I manoscritti chigiani di musica organo-cembalistica della Biblioteca apostolica vaticana', *L'organo*, v (1964–7), 63
R. Hudson: *The Development of Italian Keyboard Variations on the Passacaglio and Ciaccona from Guitar Music in the Seventeenth Century* (diss., U. of California, Los Angeles, 1967)
D. Plamenac: 'Alcune osservazioni sulla struttura del codice 117 della Biblioteca communale di Faenza', *L'ars nova italiana del trecento II: Certaldo 1969*, 161
M. Kugler: *Die Tastenmusik im Codex Faenza* (Tutzing, 1972)
A. Silbiger: *Italian Manuscript Sources of 17th-century Music* (Ann Arbor, 1980)
R. A. Hudson: *Passacaglio and Ciaccona* (Ann Arbor, 1981)

3. France

A. de La Fage: *Essais de dipthérographie musicale* (Paris, 1864/*R*1964)
E. von Werra: 'Beiträge zur Geschichte des französischen Orgelspiels', *KJb*, xxiii (1910), 37
A. Tessier: 'Une pièce d'orgue de Charles Raquet et de Mersenne de la Bibliothèque des Minimes de Paris', *RdM*, x (1929), 275
W. Apel: 'Du nouveau sur la musique française pour orgue au XVIe siècle', *ReM*, no.172 (1937), 96
N. Dufourcq: *La musique d'orgue française de Jean Titelouze à Jehan Alain* (Paris, 1941, 2/1949)
M. Reimann: *Untersuchungen zur Formgeschichte der französischen Klavier-Suite* (Regensburg, 1941)
W. Mellers: *François Couperin and the French Classical Tradition* (London, 1950/*R*1968, rev. 2/1987)
A. C. Howell jr: 'French Baroque Organ Music and the Eight Church Tones', *JAMS*, xi (1958), 106
W. Apel: 'Attaingnant: Quatorze Gaillardes', *Mf*, xiv (1961), 361
D. Fuller: *18th-century French Harpsichord Music* (diss., U. of Harvard, 1965)
D. Heartz: *Pierre Attaingnant: Royal Printer of Music* (Berkeley and Los Angeles, 1969)
N. Dufourcq: *Le livre de l'orgue français 1589–1789*, iv: *La musique* (Paris, 1972)

B. Gustafson: *French Harpsichord Music of the 17th Century: a Thematic Catalog of the Sources with Commentary*, i–iii (Ann Arbor, 1979)

J. P. Kitchen: *Harpsichord Music of 17th-century France, with Particular Emphasis on the Work of Louis Couperin* (diss., U. of Cambridge, 1979)

D. Ledbetter: *Harpsichord and Lute Music in 17th-century France* (Basingstoke, 1987)

4. Germany, Scandinavia and Eastern Europe

F. Arnold and H. Bellermann: *Das Locheimer Liederbuch nebst der Ars organisandi von Conrad Paumann* (Wiesbaden, 1864, rev. 3/1926/*R*1969)

[R. Eitner]: 'Das Buxheimer Orgelbuch', *MMg*, xx (1888), suppl.2 [incl. exx. from this MS and that of Kleber]

H. Panum: 'Melchior Schild oder Schildt', *MMg*, xx (1888), 27 [see also suppl.3, p.35, for 2 pieces]

C. Päsler: 'Fundamentbuch von Hans von Constanz', *VMw*, v (1889), 1–192

W. Nagel: 'Fundamentum Authore Johanne Buchnero', *MMg*, xxiii (1891), 71–109

M. Seiffert: 'J. P. Sweelinck und seine direkten deutschen Schüler', *VMw*, vii (1891), 145–260

J. Richter: *Katalog der Musik-Sammlung auf der Universitäts-Bibliothek in Basel* (Leipzig, 1982)

H. K. Löwenfeld: *Leonhard Kleber und seine Orgeltabulatur* (Berlin, 1897)

A. Chybiński: 'Polnische Musik und Musikkultur des 16. Jahrhunderts in ihren Beziehungen zu Deutschland', *SIMG*, xiii (1911–12), 463–505

W. Merian: *Die Tabulaturen des Organisten Hans Kotter* (Leipzig, 1916/*R*1973)

M. Grafczyński: *Über die Orgeltabulatur des Martin Leopolita* (diss., U. of Vienna, 1919)

Z. Jachimecki: 'Eine polnische Orgeltabulatur aus dem Jahre 1548', *ZMw*, ii (1919–20), 206

P. Nettl: 'Die Wiener Tanzkompositionen in der zweiten Hälfte des 17. Jahrhunderts', *SMw*, viii (1921), 45–175

H. Schnoor: 'Das Buxheimer Orgelbuch', *ZMw*, iv (1921–2), 1

A. Scheide: *Zur Geschichte des Choralvorspiels* (Hildburghausen, 1923)

C. Mahrenholz: *Samuel Scheidt: sein Leben und sein Werk* (Leipzig, 1924/*R*1968)

W. Merian: *Der Tanz in den deutschen Tabulaturbüchern* (Leipzig, 1927/*R*1968) [incl. 184 pieces from tablatures of Ammerbach, Kotter, Loeffelholz, Nörmiger, Paix and Schmid the elder]

BIBLIOGRAPHY

H. J. Moser: *Paul Hofhaimer: ein Lied- und Orgelmeister des deutschen Humanismus* (Stuttgart, 1929, rev. 2/1965) [incl. transcr. of complete works repr. from *91 gesammelte Tonsätze Paul Hofhaimers und seines Kreises* (Stuttgart, 1929)]

——: 'Eine Trienter Orgeltabulatur aus Hofhaimers Zeit', *Studien zur Musikgeschichte: Festschrift für Guido Adler* (Vienna, 1930), 84

P. Hamburger: 'Ein handschriftliches Klavierbuch aus der ersten Hälfte des 17. Jahrhunderts', *ZMw*, xiii (1930–31), 133

G. Kittler: *Geschichte des protestantischen Orgelchorals* (Ückermünde, 1931)

F. Dietrich: *Geschichte des deutschen Orgelchorals im 17. Jahrhundert* (Kassel, 1932)

F. Feldmann: 'Ein Tabulaturfragment des Breslauer Dominikaner Klosters aus der Zeit Paumanns', *ZMw*, xv (1932–3), 241

W. Apel: 'Die Tabulatur des Adam Ileborgh', *ZMw*, xvi (1933–4), 193

O. A. Baumann: *Das deutsche Lied und seine Bearbeitungen in den frühen Orgeltabulaturen* (Kassel, 1934)

J. Handschin: 'Orgelfunktionen in Frankfurt a. M. im 15. und 14. Jahrhundert', *ZMw*, xvii (1935), 108

A. Chybiński: 'Warszawska tabulatura organowa z XVII wieku', *PRM*, ii (1936), 100

L. Schrade: 'Die Messe in der Orgelmusik des 15. Jahrhunderts', *AMf*, i (1936), 129–75

W. Apel: 'Early German Keyboard Music', *MQ*, xxiii (1937), 210

F. Feldmann: 'Mittelalterliche Musik und Musikpflege in Schlesien', *Deutsches Archiv für Landes- und Volksforschung*, ii (1937)

F. Hirtler: 'Neu aufgefundene Orgelstücke von J. U. Steigleder und Johann Benn', *AMf*, ii (1937), 92

F. Feldmann: *Musik und Musikpflege im mittelalterlichen Schlesien* (Breslau, 1938/*R*1973)

G. Knoche: 'Der Organist Adam Ileborgh von Stendal: Beiträge zur Erforschung seiner Lebensumstände', *Festschrift des 600jähr. Gymnasiums zu Stendal* (Stendal, 1938); repr. in *Franziskanische Studien*, xxviii/1 (1941)

W. R. Nef: 'Der St. Galler Organist Fridolin Sicher und seine Orgeltabulatur', *Schweizerisches Jb für Musikwissenschaft*, vii (1938), 3–215

L. Schrade: 'The Organ in the Mass of the 15th Century', *MQ*, xxviii (1942), 329, 467

F. Welter: *Katalog der Musikalien der Ratsbücherei Lüneburg* (Lippstadt, 1950)

H. Federhofer: 'Eine Kärntner Orgeltabulatur', *Carinthia I*, cxlii (1952), 330

C. Mahrenholz: 'Aufgabe und Bedeutung der *Tabulatura nova*', *Musica*, viii (1954), 88

G. Most: *Die Orgeltabulatur von 1448 des Adam Ileborgh aus Stendal* (Stendal, 1954)

S. Podolsky: *The Variation Canzona for Keyboard Instruments in Italy, Austria and Southern Germany in the Seventeenth Century* (diss., Boston U., 1954)

A. Booth: *German Keyboard Music in the 15th Century* (diss., U. of Birmingham, 1954–5)

M. Reimann: 'Pasticcios und Parodien in norddeutschen Klavier-tabulaturen', *Mf*, viii (1955), 265

W. Schrammek: 'Zur Numerierung im Buxheimer Orgelbuch', *Mf*, ix (1956), 298

J. H. van der Meer: 'The Keyboard Works in the Vienna Bull Manuscript', *TVNM*, xviii/2 (1957), 72–105

J. H. Schmidt: *Johannes Buchner, Leben und Werk* (diss., U. of Freiburg, 1957)

A. Basso: 'La musica strumentale del rinascimento polacco', *RaM*, xxviii (1958), 293

K. Kotterba: *Die Orgeltabulatur des Leonhard Kleber: ein Beitrag zur Orgelmusik der ersten Hälfte des 16. Jahrhunderts* (diss., U. of Freiburg, 1958)

G. Pietzsch: 'Orgelbauer, Organisten und Orgelspiel in Deutschland bis zum Ende des 16. Jahrhunderts', *Mf*, xi (1958), 160, 307, 455; xii (1959), 25, 152, 294, 415; xiii (1960), 34

A. Osostowicz: 'Nieznany motet Diomedesa Catoni i jego utwory organowe z toruńskiej tabulatury', *Muzyka kwartalnik*, iv/3 (1959), 45

A. Sutkowski: 'Tabulatura organowa Cystersow z Pelplina', *Ruch muzyczny*, iii/1 (1959), 14

J. Gołos: 'Zaginiona tabulatura organowa Warszawskiego towarzystwa muzycznego (ca. 1580)', *Muzyka kwartalnik*, v/4 (1960), 70

F. W. Riedel: *Quellenkundliche Beiträge zur Geschichte der Musik für Tasteninstrumente in der 2. Hälfte des 17. Jahrhunderts* (Kassel and Basle, 1960)

H. Schmid: 'Una nuova fonte di musica organistica del secolo xvii', *L'organo*, i (1960), 107

A. Sutkowski: 'Nieznane polonika muzyczne z XVI i XVII wieku', *Muzyka kwartalnik*, v/1 (1960), 62

T. Göllner: *Formen früher Mehrstimmigkeit in deutschen Handschriften des späten Mittelalters* (Tutzing, 1961)

G. Gołos: 'Il manoscritto I/220 della Società di musica di Varsavia, importante fonte di musica organistica cinquecentesca', *L'organo*, ii (1961), 129

J. Gołos: 'Tabulatura Warsawskiego towarzystwa musycznego jako źródto muzyki organowej', *Muzyka kwartalnik*, vi/4 (1961), 60

B. Lundgren: 'Nikolajorganisten Johan Lorentz i Köpenhamn', *STMf*, xliii (1961), 249

L. Schierning: *Die Überlieferung der deutschen Orgel- und Klaviermusik aus der 1. Hälfte des 17. Jahrhunderts* (Kassel and Basle, 1961)
A. Sutkowski and O. Mischiati: 'Una preciosa fonte di musica strumentale: l'intavolatura di Pelplin', *L'organo*, ii (1961), 53
W. Apel: 'Neu aufgefundene Clavierwerke von Scheidemann, Tunder, Froberger, Reincken und Buxtehude', *AcM*, xxxiv (1962), 65
T. Dart: 'Elisabeth Eysbock's Keyboard Book', *STMf*, xliv (1962), 5; also in *Hans Albrecht in memoriam* (Kassel, 1962), 84
G. S. Gołos: 'Tre intavolature manoscritte di musica vocale rintracciate in Polonia', *L'organo*, iii (1962), 123
A. Reichling: 'Die Präambeln der Hs. Erlangen 554 und ihre Beziehungen zur Sammlung Ileborghs', *GfMKB, Kassel 1962*, 109
L. Schierning: *Quellengeschichtliche Studien zur Orgel- und Klaviermusik in Deutschland aus der 1. Hälfte des 17. Jahrhunderts* (Kassel and Basle, 1962)
H. J. Marx: 'Der Tabulatur-Codex des Basler Humanisten Bonifacius Amerbach', *Musik und Geschichte: Leo Schrade zum 60. Geburtstag* (Cologne, 1963), 50
O. Mischiati: 'L'intavolatura d'organo tedesca della Biblioteca nazionale di Torino, *L'organo*, iv (1963), 1–154
F. W. Riedel: *Das Musikarchiv im Minoritenkonvent zu Wien*, CaM, i (1963)
E. Southern: *The Buxheim Organ Book* (Brooklyn, 1963)
—— : 'Some Keyboard Basse Dances of the Fifteenth Century', *AcM*, xxv (1963), 114
A. Sutkowski and A. Osostowicz-Sutkowska: *The Pelplin Tablature: a Thematic Catalogue*, AMP, i (1963)
J. R. White: 'The Tablature of Johannes of Lublin', *MD*, xvii (1963), 137
H. R. Zöbeley: *Die Musik des Buxheimer Orgelbuchs: Spielvorgang, Niederschrift, Herkunft, Faktur* (Tutzing, 1964)
F. Blume: *Geschichte der evangelischen Kirchenmusik* (Kassel, 2/1965; Eng. trans., enlarged, 1974 as *Protestant Church Music: a History*)
F. Crane: '15th-century Keyboard Music in Vienna MS 5094', *JAMS*, xviii (1965), 237
W. Apel: 'Die Celler Orgeltabulatur von 1601', *Mf*, xix (1966), 142
W. Breig: *Die Orgelwerke von Heinrich Scheidemann* (Wiesbaden, 1967)
T. Göllner: 'Notationsfragmente aus einer Organistenwerkstatt des 15. Jahrhunderts', *AMw*, xxiv (1967), 170
—— : 'Eine Spielanweisung für Tasteninstrumente aus dem 15. Jahrhundert', *Essays in Musicology: a Birthday Offering for Willi Apel* (Bloomington, 1968), 69
G. Leonhardt: 'Johann Jakob Froberger and his Music', *L'organo*, vi (1968), 15
M. Schuler: 'Eine neu entdeckte Komposition von Adam Steigleder', *Mf*, xxi (1968), 42

C. Wolff: 'Conrad Paumanns Fundamentum organisandi und seine verschiedenen Fassungen', *AMw*, xxv (1968), 196

W. Apel: 'Der deutsche Orgelchorale um 1600', *Musa–mens–musici: im Gedenken an Walther Vetter* (Leipzig, 1969), 67

T. Göllner: 'Die Trecentonotation und der Tactus in den ältesten deutschen Orgelquellen', *L'ars nova italiana del trecento II: Certaldo 1969*, 176

C. Wolff: 'Arten der Mensuralnotation im 15. Jahrhundert und die Anfänge der Orgeltabulatur', *GfMKB, Bonn 1970*, 609

J. Stenzl: 'Un' intavolatura tedesca sconosciuta della prima metà del cinquecento', *L'organo*, x (1972), 51–82

S. Wollenberg: *Viennese Keyboard Music in the Reign of Karl VI (1712–40): Gottlieb Muffat and his Contemporaries* (diss., U. of Oxford, 1975)

5. Low Countries

C. van den Borren: *Les origines de la musique de clavier dans les Pays-Bas (nord et sud) jusque vers 1630* (Brussels, 1914)

—— : 'Le livre de clavier de Vincentius de la Faille (1625)', *Mélanges de musicologie offerts à M. Lionel de La Laurencie* (Paris, 1933), 85

S. Clercx: 'Les clavecinistes belges', *ReM*, no.192 (1939), 11

A. E. F. Dickinson: 'A Forgotten Collection: a Survey of the Weckmann Books', *MR*, xvii (1956), 97

T. Dart: 'John Bull's "Chapel"', *ML*, xl (1959), 279

W. Breig: 'Der Umfang des choralgebundenen Orgelwerkes von Jan Pieterszoon Sweelinck', *AMw*, xvii (1960), 258

A. Curtis: Introduction to *Nederlandse klaviermuziek uit de 16e en 17e eeuw*, MMN, iii (1961)

T. Dart: 'The Organ-book of the Crutched Friars of Liège', *RBM*, xvii (1963), 21

A. E. F. Dickinson: 'The Lübbenau Keyboard Books: a Further Note on Faceless Features', *MR*, xxvii (1966), 270

W. Breig: 'Die Lübbenauer Tabulaturen Lynar A1 und A2: eine quellenkundliche Studie', *AMw*, xxv (1968), 96, 223

A. Curtis: *Sweelinck's Keyboard Works: a Study of English Elements in Seventeenth-century Dutch Composition* (London and Leiden, 1969, 2/1972)

6. Spain and Portugal

H. Anglès: 'Orgelmusik der Schola Hispanica vom XV. bis XVII. Jahrhundert', *Festschrift Peter Wagner* (Leipzig, 1926/*R*1969), 11

W. Apel: 'Early Spanish Music for Lute and Keyboard Instruments', *MQ*, xx (1934), 289

M. S. Kastner: *Música hispânica: o estilo musical de Padre Manuel R. Coelho* (Lisbon, 1936)

—— : *Contribución al estudio de la música española y portuguesa* (Lisbon, 1941)

—— : 'Los manoscritos musicales ns. 48 y 242 de la Biblioteca Geral de la Universidad de Coimbra', *AnM*, v (1950), 78

J. Moll Roqueta: 'Músicos de la corte dal Cardenal Juan Tavera (1523–1545): Luis Venegas de Henestrosa', *AnM*, vi (1951), 156

K. Speer: 'The Organ *Verso* in Iberian Music to 1700', *JAMS*, xi (1958), 189

B. Hudson: *A Portuguese Source of Seventeenth-century Iberian Organ Music* (diss., Indiana U., 1961)

H. Anglès: 'Die Instrumentalmusik bis zum 16. Jahrhundert in Spanien', *Natalicia musicologica Knud Jeppesen* (Copenhagen, 1962), 143

W. Apel: 'Spanish Organ Music of the Early 17th Century', *JAMS*, xv (1962), 174

A. C. Howell jr: 'Cabezón: an Essay in Structural Analysis', *MQ*, l (1964), 18

R. A. Hudson: *Passacaglio and Ciaccona* (Ann Arbor, 1981)

7. *Great Britain*

J. A. Fuller Maitland and A. H. Mann: *Catalogue of Music in the Fitzwilliam Museum, Cambridge* (London, 1893)

J. Wolf: 'Zur Geschichte der Orgelmusik im vierzehnten Jahrhundert', *KJb*, xiv (1899), 14

E. W. Naylor: *An Elizabethan Virginal Book* (London, 1905)

E. Walker: *A History of Music in England* (Oxford, 1907, rev. 3/1952 by J. A. Westrup)

J. E. West: 'Old English Organ Music', *PMA*, xxxvii (1910–11), 1

C. van den Borren: *Les origines de la musique de clavier en Angleterre à l'époque de la Renaissance* (Brussels, 1913; Eng. trans., 1913)

W. Niemann: *Die Virginalmusik* (Leipzig, 1919)

M. Glyn: *About Elizabethan Virginal Music and its Composers* (London, 1924, enlarged 2/1934)

M.-L. Pereyra: 'Les livres de virginal de la bibliothèque du Conservatoire de Paris', *RdM*, vii (1926), 204; viii (1927), 36, 205; ix (1928), 235; x (1929), 32; xii (1931), 22; xiii (1932), 86; xiv (1933), 24

L. Neudenberger: *Die Variationstechnik der Virginalisten im Fitzwilliam Virginal Book* (Berlin, 1937)

H. M. Miller: 'Sixteenth-century English Faburden Compositions for Keyboard', *MQ*, xxvi (1940), 50

—— : 'The Earliest Keyboard Duets', *MQ*, xxix (1943), 438

—— : *English Plainsong Composition for Keyboard in the Sixteenth Century* (diss., Harvard U., 1943)

—— : 'Pretty Wayes: for young Beginners to Looke on', *MQ*, xxxiii (1947), 543

R. Donington and T. Dart: 'The Origin of the English In Nomine', *ML*, xxx (1949), 101
E. H. Fellowes: 'My Ladye Nevells Book', *ML*, xxx (1949), 1
H. M. Miller: 'Fulgens Praeclara: a Unique Keyboard Setting of a Plainsong Sequence', *JAMS*, ii (1949), 97
G. Reese: 'The Origins of the English "In Nomine"', *JAMS*, ii (1949), 7
F. Dawes: 'Nicholas Carlton and the Earliest Keyboard Duet', *MT*, xcii (1951), 542
J. Ward: 'The "Dolfull Domps"', *JAMS*, iv (1951), 111
D. Stevens: *The Mulliner Book: a Commentary* (London, 1952)
—— : 'A Unique Tudor Organ Mass', *MD*, vi (1952), 167
E. E. Lowinsky: 'English Organ Music of the Renaissance', *MQ*, xxxix (1953), 373, 528
T. Dart: 'Le manuscrit pour le virginal de Trinity College, Dublin', *La musique instrumentale de la Renaissance: CNRS Paris 1954*, 237
—— : 'A New Source of Early English Organ Music', *ML*, xxxv (1954), 201
—— : 'New Sources of Virginal Music', *ML*, xxxv (1954), 93
J. Jacquot: 'Sur quelques formes de la musique de clavier élisabéthaine', *La musique instrumentale de la Renaissance: CNRS Paris 1954*, 241
J. Ward: 'Les sources de la musique pour le clavier en Angleterre', *La musique instrumentale de la Renaissance: CNRS Paris 1954*, 225
J. L. Boston: 'Priscilla Bunbury's Virginal Book', *ML*, xxxvi (1955), 365
H. M. Miller: 'Forty Wayes of 2 Pts. in One of Tho[mas] Woodson', *JAMS*, viii (1955), 14
D. Stevens: 'Further Light on *Fulgens praeclara*', *JAMS*, ix (1956), 1
F. Ll. Harrison: *Music in Medieval Britain* (London, 1958, 2/1963)
H. J. Steele: *English Organs and Organ Music from 1500 to 1650* (diss., U. of Cambridge, 1958)
R. L. Adams: *The Development of Keyboard Music in England during the English Renaissance* (diss., U. of Washington, 1960)
J. Stevens: *Music and Poetry in the Early Tudor Court* (London, 1961)
H. Ferguson: 'Repeats and Final Bars in the Fitzwilliam Virginal Book', *ML*, xliii (1962), 345
E. Apfel: 'Ostinato und Kompositionstechnik bei den englischen Virginalisten der elisabethanischen Zeit', *AMw*, xix–xx (1962–3), 29
T. Dart: 'Notes on a Bible of Evesham Abbey (ii): a Note on the Music', *English Historical Review*, lxxix (1964), 777
J. A. Caldwell: *British Museum Additional Manuscript 29996: Transcription and Commentary* (diss., U. of Oxford, 1965)
—— : 'Keyboard Plainsong Settings in England, 1500–1660', *MD*, xix (1965), 129
P. le Huray: *Music and the Reformation in England, 1549–1660* (London, 1967)

G. Sargent: *A Study and Transcription of Ms. Brit. Mus. Add. 10337* (diss., Indiana U., 1968)

A. Brown: '"My Lady Nevell's Book" as a Source of Byrd's Keyboard Music', *PRMA*, xcv (1968–9), 29

G. Beechey: 'A New Source of 17th Century Keyboard Music', *ML*, l (1969), 278

A. Curtis: *Sweelinck's Keyboard Works: a Study of English Elements in Seventeenth-century Dutch Composition* (London and Leiden, 1969, 2/1972)

M. C. Maas: *Seventeenth-century English Keyboard Music: a Study of Manuscripts Rés. 1185, 1186 and 1186bis of the Paris Conservatory Library* (diss., Yale U., 1969)

J. Caldwell: 'The Pitch of Early Tudor Organ Music', *ML*, li (1970), 156

T. Dart: 'An Early Seventeenth-century Book of English Organ Music for the Roman Rite', *ML*, lii (1971), 27

B. A. R. Cooper: 'Albertus Bryne's Keyboard Music', *MT*, cxiii (1972), 142

—— : 'The Keyboard Suite in England before the Restoration', *ML*, liii (1972), 309

M. Boyd: 'Music MSS in the Mackworth Collection at Cardiff', *ML*, liv (1973), 133

J. Caldwell: *English Keyboard Music Before the Nineteenth Century* (Oxford, 1973)

F. Routh: *Early English Organ Music from the Middle Ages to 1837* (London, 1973)

M. Tilmouth: 'York Minster MS M.16(s) and Captain Prendcourt', *ML*, liv (1973), 302

B. A. R. Cooper: *English Solo Keyboard Music of the Middle and Late Baroque* (diss., U. of Oxford, 1974)

N. Bergenfeld: *The Keyboard Fantasy of the Elizabethan Renaissance* (diss., New York U., 1978)

O. Neighbour: *The Consort and Keyboard Music of William Byrd* (London, 1978)

R. Petrie: 'A New Piece by Henry Purcell', *Early Music*, vi (1978), 374

B. A. R. Cooper: 'Keyboard Sources in Hereford', *RMARC*, xvi (1980), 135

Index

INDEX

INDEX

INDEX

INDEX